EXTRAORDINARY ENGINEERS

Volume 3

DR. J. A. SANCHEZ
& THE
EXTRAORDINARY ENGINEERS

Extraordinary Engineers
By Dr. J. A. Sanchez

Extraordinary Engineers:
Extraordinary Engineers Volume 3
Female Engineers of This Day and Time
Radiant Pearls Publishing
2026

All journeys in this book are contributions of the extraordinary engineers.
Printed in the U.S.A.
ISBN: 979-8-9958883-0-7 (color hardcover)
ISBN: 979-8-9958883-1-4 (color paperback)
ISBN: 979-8-9958883-2-1 (black & white paperback)
ISBN: 979-8-9958883-3-8 (ebook)

Cover art done by Get Covers https://getcovers.com/
Editing done by Sandra the Editor

Contents

Introduction
The Start of Extraordinary Engineers

It was the summer of 2022, and I had just been a part of a STEM Summit. I was incredibly inspired by the other women on the panel. For days, my heart and mind were stirring, knowing I needed to do something. I also knew when God calls you to something, you do it. I had a simple but powerful vision: I had to capture their stories! Stories of women who could inspire a generation for years to come, just as I had been inspired!

I didn't grow up seeing engineers who looked like me, and I didn't fully understand what engineering was or what it could become in my life. My path into this field was not traditional, and at times, it felt uncertain. But what I've learned along the way is that there is no one path. And that is exactly what makes this journey so powerful.

Now, as I bring you Volume 3 of Extraordinary Engineers: Female Engineers of This Day and Time, I am reminded of why this work matters. This is the very resource I needed 20 years ago.

The women in this book each have their own story. Different beginnings, different challenges, different definitions of success. Yet they all share something in common: a willingness to keep going, to grow, and to step into spaces where they may not have always seen themselves. And to bring others with them along the way.

This book is more than a collection of stories. It is a reflection of what is possible.

Through these pages, you'll see real journeys. The highs, the struggles, the pivots, and the defining moments. You'll see that success doesn't look one way, and that meaningful careers and lives are often built through unexpected opportunities and perseverance.

What started as an idea has become a platform for visibility, inspiration, and connection. This volume continues that momentum.

My hope is that as you read these stories, you see a piece of yourself in them. That you grab onto that thing that connects you, and you run with it. I want you to be inspired, encouraged, and reminded that your path, no matter how different it may look, is your very own, and there is power in that!

Because the truth is, there is no single way to become an engineer. There is no single way to live an extraordinary life.

To the next generation, to those already on the journey, and to anyone ready to take the next step forward, this book is for you!

With Love,

Justina Sanchez

Matthew 6:33 "Seek first the Kingdom of God and His righteousness, and all these things shall be added to you."

EXTRAORDINARY ENGINEERS
VOLUME 3

Extraordinary Engineers

Cleared for Takeoff: Building a Life of Purpose, Resilience, and Flight

Coauthored By Eileen M. Vélez-Vega, PE. (PR, FL, USVI)

Former Puerto Rico Secretary of the Department of Transportation & Public Works

I was born in Sabana Grande, el pueblo del Petate, a small town nestled in the southwest of Puerto Rico. It's the kind of place where everyone knows your name, especially when your mom is one of the town's beloved English teachers, and where dreams often feel bigger than the town itself. My parents, Eileen and Sergio, despite limited financial resources, always told my brother Omar and me that there were no limits to what I could accomplish. But they also made it clear: "You must work hard and ultimately, ultimately, it's up to you to make it happen!" I had to work hard and never forget to pray! Simple concepts, right!

That message became my mantra.

My love for STEM began in the eighth grade. My teachers saw something in me and filled out an application for a STEM camp at the InterAmerican University in San Germán. I didn't know what I was signing up for until I was accepted. That experience opened a door I didn't even know existed. Then, not long after, a teacher in our town organized a trip to Space Camp at the Kennedy Space Center in Cape Canaveral, Florida. I wanted to go more than anything, but my parents couldn't afford it. Still, they never said no. That's when I learned about fundraisers. My mom and dad even sold "bacalaitos" at the town's Christmas festival. I sold chocolates, bread, and cakes around my neighborhood; everyone was very supportive and helped us raise the money. And we did it. I made it to Space Camp, and I fell in love with space, with science, with the idea that maybe—just maybe—I could become an astronaut. I still haven't let go of that dream.

Even though I hadn't traveled much, I dreamed of seeing the world. That dream started to take shape in high school, in tenth grade, when I finally got to travel to Space Camp. After that, there was no stopping me. I won a lottery for academic scholars from the Government of Puerto Rico and spent ten unforgettable days in Mexico and Dallas with other public-school students from across the island. It was my first experience with international travel, and I loved every second of it.

To help fund another opportunity—Close Up in Washington, D.C.—I worked at the local Loctite factory in the Human Resources department as part of a special student program. But I wasn't just about science and engineering. I also loved dancing and public speaking. So,

I auditioned for the dance team at the Center for Performing Arts (CABA, for its Spanish acronym), part of a special program through the Puerto Rico Department of Education. In the summer of 1998, right before starting my freshman year in college, I was chosen as part of a delegation to represent Puerto Rico in international dance festivals in Belgium and Italy. I spent 49 days dancing folkloric ballet, meeting people from all over the world, and sharing our culture on global stages. I was just a teenager, but I felt like an ambassador.

I arrived at college two weeks after classes had already begun. My friends thought I was crazy. "You're going to flunk," they warned me. But I didn't. I caught up, I kept up, and I never regretted taking that leap. That international experience had changed me, and I knew there was a bigger world out there, and I wanted to see it.

When I started college, my world shifted. I was a freshman when my parents divorced. It was a difficult moment, but even then, our family bond remained unshaken. I never felt alone or unsupported. My mother, especially, had always been my anchor and biggest cheerleader. She had always dreamed of becoming a flight attendant in the 1970s, but her family couldn't support her. But she never let go of her dreams entirely; she passed them on to me. She is the reason I fell in love with aviation. She is my inspiration.

I studied civil engineering at the University of Puerto Rico at Mayagüez, "El Colegio". During my undergraduate years, I interned at Walt Disney World in Orlando, Florida and NASA's Marshall Space Flight Center in Huntsville, Alabama.

At Walt Disney World, I was part of the College Program. I was placed in the Cast and Entertainment Department at Animal Kingdom, and I absolutely loved it. I was Mickey and Minnie's friend, as well as Meeko and Br'er Rabbit's. Every day, I got to bring joy to families from around the world. It was magical, yes, but I was also a civil engineering student, and I wanted to know what kind of magic the engineers were creating behind the scenes.

One day, I gathered the courage to ask my supervisor if I could meet a Disney engineer. I didn't know what would come of it, but I believed in asking. To my surprise, he said yes, and he made it happen. I shadowed the Chief of Structural and Design at Disney. That day changed everything for me. I saw firsthand what Disney engineers—Imagineers—really did. They were building structures and experiences, dreams, and stories. I was completely captivated.

Looking back, I realize how powerful that moment was—not just because of where it led me, but because it taught me something I carry with me to this day: never underestimate the power of self-initiative. Be brave enough to ask. Be bold enough to dream even when you haven't seen anyone do it or don't know how to make it happen yet. Magic helpers are real—but they often show up when you take the first step.

I left Disney inspired and determined to aim even higher. So, I did what I always do when something excites me: I took initiative. The following summer, I applied to the NASA Undergraduate Student Research Program (USRP). I still remember the day I saw the announcement on a bulletin board in the electrical engineering department —back then, in 2001, we didn't use the internet or email the way we do now. The deadline was getting close, and I didn't have time to overthink it. I ran—literally—to the registrar's office to get my official transcript, tracked down my professors and asked for letters of recommendation. I remember mailing my application with a mix of hope and nerves. I mailed the envelope just in time, making sure it was postmarked by the deadline. And then I waited.

A few months later, in the spring of 2001, the letter arrived. I opened it. I had been accepted to NASA. I couldn't believe it. There was a lot of screaming, and probably some tears too—definitely from my mom and me. It was one of those moments that stays with you forever. A moment when the world opens up just a little wider, and you realize that your dreams are not only valid—they're possible.

That summer, I joined a group of 25 students from across the USA at NASA's Marshall Space Flight Center. Out of the entire cohort, only two of us were from Puerto Rico. I remember feeling both proud and humbled to represent my island in such a prestigious program. Walking through the gates of NASA felt like stepping into a dream. I worked in the Advanced Projects Office as part of the Flights Directorate, and I was working on real engineering challenges, contributing to projects that had national and even global impact.

During the internship, I attended several workshops and events that were offered to us as part of the Program. Most of these sessions were held in the former 4200 building, a place that quickly became familiar and exciting. We had the opportunity to meet astronauts who had just returned from space. It felt surreal to be in the same room with them, and I just hoped that would someday be me. The workshops were good, but one in particular left a mark on me that I carry to this day.

The workshop was different. It wasn't about rockets or propulsion systems or the latest research. It was about people, about diversity, and the role of minorities in the NASA space program. The speaker was an African American woman—an incredible force of energy, motivation, and pride. She spoke with such confidence and passion about her work at NASA, and about what it meant to be a minority. She made it clear that we all belonged, that our voices and perspectives mattered, and that our presence was part of NASA's strength. I was inspired, but I didn't think she was talking to me.

Later that evening, back at the dorms at the University of Alabama in Huntsville, I was talking with my roommate about the workshop. And she said, "Oh, honey, you are very much a minority." My reaction was "Oh… cool. That's great," but a bit shocked.

It hadn't occurred to me that I was not a minority in Puerto Rico because we were all Puerto Rican, and nobody ever talked about us being different or underrepresented or marginalized; those words did not exist in my language. I hadn't yet understood what it meant to be a minority in a professional space. That was a quiet but powerful realization, and it never felt like an obstacle—it felt like I was special and with some kind of superpower. Being a Latina is a superpower! Instead of feeling discouraged, I felt proud. That NASA speaker had made such an impression on me that I was happy to be as special as she was. I wanted to carry myself with the same confidence, the same pride, the same sense of purpose. She even recommended a book during her talk, "Who Moved My Cheese?" by Spencer Johnson, M.D. I checked it out from the library that same week. It was a simple book, but its message about change, adaptability, and courage resonated deeply with me. To this day, I recommend it in every STEM presentation I give, whether

I'm speaking to students or professionals. It's a reminder that change is inevitable, but growth is a choice.

I wish I could talk with that NASA speaker again and let her know how powerful her presentation has been in my life! I did see her one more time exiting an elevator at NASA and I told her I had read the book and she was so excited that she told her co-workers, which also made me feel good and taught me the importance of providing feedback and having the courage to talk with people that might seem a bit intimidating, because they might just be accessible and kind.

Before the summer ended, I knew I wasn't ready to leave. I had built such strong relationships with my co-workers in the Advanced Projects Office—my office buddy, Mr. Wayne, my supervisor, David, and the chief, Joe (may he rest in peace and power), had all become mentors and friends. And again, I did what I always do when something feels right: I spoke up. To my surprise and joy, they offered me a full Co-op internship. I extended my stay through December 2001, and then I returned the following summer. NASA wasn't just a summer internship anymore; it had become a second home.

What made it even more special was the community I found within NASA. I met so many inspiring Puerto Ricans working there—engineers, scientists, professionals who welcomed me like family. They looked out for me, mentored me, and made sure I was safe and supported. One of those special fellow Puerto Ricans was Monsi Román. Monsi became like an aunt to me. She took me under her wing, and to this day—24 years later—she's still an important part of my life. Later in my story, I'll tell you how she changed my life in ways I never expected.

Thinking back on that summer at NASA, it wasn't just about research or engineering. It became much deeper. It was about identity. It was about representation. I arrived in Huntsville as a curious and nervous engineering student from Puerto Rico, and I left more confident, more determined, and more committed to opening doors for others like me. It also provided me with credibility and respect.

Looking back, I can say with certainty: Disney magic helped me dream, and NASA helped me believe. And it all started because I was brave enough to ask a question, to take a chance, and to believe that I belonged in those rooms—even when no one else looked like me.

When I returned to college after my NASA internship, I was still deeply focused on my studies, but I was also starting to notice something beyond books and engineering formulas—I was noticing life. And part of that life was a classmate I had known since freshman year. His name was Airangel. He had always been a nice guy, very handsome, kind, and suddenly very interesting.

Airangel was part of the ASCE Concrete Canoe Team, and after a serious accident involving his Achilles tendon, I started helping him with his books and studying. What began as a friendship turned into something more. We started dating, and 23 years later, we're both civil engineers, married since 2005, and the proud parents of a beautiful, smart, and kind daughter named Anna Isabelle. She's now 14, and every day she reminds me that life is about so much more than just books and degrees. It's about love, family, and the people who walk beside you on the journey.

In college, I also became extremely involved in the Society of Women Engineers (SWE). I attended my first SWE national conference, surrounded by thousands of women engineers—it was eye-opening and made me feel part of a powerful sisterhood. I was the chair of an SWE regional conference at our university, which gave me the chance to meet so many SWE leaders and pioneers. Their stories, their strength, and their presence all inspired me to keep going, to keep growing, and to become a stronger leader. At the same time, I was active in the American Society of Civil Engineers (ASCE). I joined the Steel Bridge Team, and in 2003, for the first time ever, our university won first place at the Regional Student Championships. That win gave us a huge boost of confidence as young engineers preparing to graduate. Currently, both SWE and ASCE are a big part of my career journey.

I began my professional career in 2003 with the U.S. Army Corps of Engineers at the Engineer Research and Development Center (ERDC) in Vicksburg, Mississippi. I was a research civil engineer in the Airfields and Pavements Branch (APB) within the Geotechnical and Structures Laboratory. It was a highly technical role, and I was part of a team developing rapid construction technologies for military airfield projects. My ERDC APB colleagues became my friends, and to this day, they are a team I admire deeply. I'm forever grateful to them for teaching me so much and for the camaraderie we built through the years.

Moving to Mississippi was tough. I missed home, the culture, the warmth of Puerto Rico. But the work was incredible. I collaborated with different military branches on accelerated pavement testing research and had the opportunity to travel the world. My favorite assignments were the Pavement Management Inspections of military airfields and a special project for the U.S. Air Force on concrete pavement rubblization. That project took me to Grand Forks, North Dakota, where I worked at the Grand Forks Air Force Base on their runway rehabilitation project. Everything was different—the weather, the landscape, the pace. I found myself working with non-destructive testing equipment, attending pre-construction meetings with military personnel, and navigating life on my own, far from home.

During that same period, I also had the incredible opportunity to work at the FAA Technical Center in New Jersey and present my rubblization research at a technical conference in Seoul, South Korea. Those experiences expanded my technical expertise and allowed me to grow a national and international network of amazing colleagues who continue to inspire and support me to this day.

Those years as a research civil engineer laid the strongest foundation I could have asked for in my career. I also pursued my master's in engineering through the ERDC Graduate Institute, a satellite campus of Mississippi State University, focusing on pavements and geotechnical engineering. It was during this time that I discovered the Transportation Research Board (TRB) and its technical committees, and I became a published technical paper author, something I never imagined when I first started my career.

In 2006, after a few unforgettable years in Mississippi, my husband and I made a big decision. As a young married couple, we knew we had to find a balance. We compromised—something we've done many times since—and found a new opportunity that brought us closer to home. That's how we landed in West Palm Beach, Florida, where we joined Kimley-Horn and Associates, Inc. The joke was I had gone to the dark side, consulting! At the time, the firm was growing fast, and it felt like the right place to take the next step in my career.

I became an analyst and then a project manager in the Aviation team, and I truly fell in love with airports. Designing runways, taxiways, and aprons became more than just technical work—it became a way to connect people, to build resilience, and to serve communities. I also loved working with my clients and learning how to network and market new work while building very strong relationships. I worked on projects across Florida, from Fort Lauderdale to Tallahassee, and every assignment deepened my passion for aviation infrastructure.

However, from the moment I started at Kimley-Horn, I knew I wanted to bring my work closer to Puerto Rico. I told my new supervisor, Dave, on day one: "I want to work in Puerto Rico." He said, "Find a client and a project, and we'll go." It took me eight years, but I never let go of that goal.

In 2010, we opened a Kimley-Horn office in Puerto Rico, and by 2014, I relocated to the island to lead and grow it. It was one of the hardest tasks of my career. I was thrilled to be back home, but Puerto Rico was in crisis—facing bankruptcy, political instability, and economic uncertainty. Many people said we were crazy to go back, but we believed Puerto Rico was worth the risk, and we wanted to give back to the island. During the worst times, I remember asking myself, "What have I done?" There wasn't much work on the island, and the first few years were incredibly challenging, but we didn't give up.

Then came Hurricane Maria in 2017. A Category 5 storm that devastated our island and changed our lives forever. To this day, we divide our lives into "before and after" Maria. It was the saddest, most impactful experience we've lived through. But it also marked a turning point; suddenly, we were rebuilding. I became the project manager for all 10 airports on the island. I worked closely with the Puerto Rico Highway and Transportation Authority on landside recovery projects and supported private retail clients with structural assessments so they could reopen and serve their communities.

It all turned around. I was at the top of my game—thriving in Puerto Rico's engineering industry. I had expanded my work to include airports in the U.S. Virgin Islands, and I finally felt like a well-rounded professional, ready to take on even more. I was presenting my aviation work at conferences across the island and the mainland, winning new projects, and building a name for myself. When you do your work with passion, people notice, and my name started to get around.

I also became more involved in the community, volunteering with local nonprofits and building a strong, stable network on the island. The recognitions started coming in, and while I never expected them, they were humbling and deeply encouraging. Airport Business magazine named me one of their Top 40 Under 40 professionals. The Civil Engineering Institute of the College of Engineers and Land Surveyors of Puerto Rico honored me as Distinguished Engineer of the Year—the first time in 80 years that a woman had received that award. SWE recognized me with the New Distinguished Engineer and Emerging Leader awards. Those honors gave me confidence, but more importantly, they gave me purpose. I knew I had to give back. I became an advocate for Latinas in STEM, working with students from pre-K all the way through college. I volunteered at math competitions, visited schools, and shared my story—especially with young girls—about what it's like to be an aviation engineer. I wanted them to see that they could do it too.

Let's take a pause and go back to 2010, because life had other plans.

While I was five months pregnant with my daughter, I was diagnosed with Hodgkin's Lymphoma. I was 30 years old, building a career I loved, and suddenly everything stopped. I underwent three rounds of chemotherapy while pregnant, something I never imagined I'd have to face. I was terrified and held on to hope, to faith, and to the strength I didn't know I had. My mom and my husband were my caregivers. And then, in December, Anna Isabelle was born—healthy, beautiful, and with a full head of hair. She carried me through many more months of chemotherapy and recovery.

BUT in 2012, I relapsed. This time, I needed a stem cell transplant. It was one of the hardest periods of my life—but also one of the most transformative. I learned what it means to fight, literally relearning how to walk and talk again. I leaned into my faith even though I was scared and clinically depressed. It took me years to recover physically and mentally, and my medical teams at Moffitt Cancer Center and the Palm Beach Cancer Institute became my angels. Anna Isabelle was my reason, and she gave me strength. After my transplant, I became a half-marathon athlete and a volunteer with the Leukemia & Lymphoma Society, raising thousands of dollars for cancer research, and helping other patients and caregivers. My journey with cancer didn't end with survival—it became a mission to help others find their strength, too.

Fast forward to 2020, we went into lockdown for nine months, but we kept working—every single one of our airport design projects stayed alive. My team of seven engineers grew to more than 60, collaborating across Kimley-Horn offices and time zones. It was a testament to our leadership, technical expertise, and commitment. I was burned out. Still, we delivered. We finished our projects on time and within budget, using innovative techniques that were later recognized in Airport Magazine. My team deserved that recognition. Through it all, I met even more incredible people within Kimley-Horn—people who inspired and reminded me why I loved my work.

Then, just before Thanksgiving 2020, I got a call that changed everything. A dear colleague I hadn't spoken to in over a year asked me for my resume to be considered for the role of Secretary of the Department of Transportation and Public Works of Puerto Rico (DTOP, for its Spanish acronym) under the new Governor, the Honorable Pedro Pierluisi. WHAT? She had to be joking. I said no. The media scrutiny, the politics, my family, my Kimley-Horn office, and my team—I couldn't do that. But she reminded me of something I had said to her years earlier, after one of my presentations at an engineering conference: "If I ever worked in government, I'd want to be Secretary of Transportation." She was holding me accountable to my own words.

Ironically, I had told my mother the same thing when I was a little girl. Be careful what you put out into the universe—it just might happen.

I sent her my resume, and minutes later, I got a call with an interview date and time. I had a Zoom interview with a four-person committee, and they forwarded my resume to the Governor. When I met him, I had already reviewed his transportation platform. I told him what I believed were the most critical issues: **maintenance and planning**. He offered me the job. I went home and talked to my husband. Like good engineers, we made an Excel spreadsheet. The job paid nearly a third of my current compensation. It was a huge risk, and I had NEVER been involved in politics. I accepted the role.

On December 7, 2020, I was designated the first woman to serve as Secretary of the Puerto Rico Department of Transportation and Public Works. In its 105-year history, there had never been a woman in that role. I was the first, but I knew I wouldn't be the last.

My life changed overnight. I became a public figure. People were Googling me, trying to find out who I was—and if they could dig up any dirt, there was none. But that didn't mean the role didn't come with hesitation and scrutiny. It was a historic moment; it was a responsibility. I focused on rebuilding after hurricanes, improving traffic safety and mobility, refocusing on Complete Streets, and maintaining the infrastructure that connects our island.

I got to work. We had over $600 million in FEMA funding for Hurricane Maria recovery, but no internal structure to manage it, and I created a federal grants office within DTOP, which helped us obligate 90% of those funds and launch 157 recovery projects. Maintenance had been neglected for decades due to lack of funding. I was able to change that. I had a small and close team, and we restructured funding through Law 22 traffic fines, securing 70% of those revenues—about $120 million annually—for road maintenance. We worked closely with FHWA and FTA to align our projects with federal priorities and unlock additional infrastructure funds through the Infrastructure Investment and Jobs Act. It was the biggest undertaking the Department had ever seen, and I was so proud to be part of it.

I also had the honor of advising the U.S. Secretary of Transportation Honorable Pete Buttigieg as part of the Transforming Transportation Advisory Committee, and served as Chair of the Aviation Council for AASHTO, and worked with leaders across the country to shape the future of transportation.

I'm an optimist by nature, but being a political appointee tested my self-confidence and emotional well-being in ways I never expected. The honor of being the first woman Secretary allowed me to inspire girls—including my own daughter—to believe that women can do anything. I was blessed to have served, and despite the negative social media, the public scrutiny, and the internal conflicts, I never compromised my integrity.

One of the biggest lessons was just how important communication, honesty, and loyalty are—and how rare they can be in politics. I had to learn quickly how to speak to the media, how to stay composed under pressure, and how to deliver clear, honest messages even when the topics were difficult. I was on TV, radio, and the printed press almost every week. It was exhausting, but it was also a powerful platform to inform the public and advocate for the work we were doing.

I completed my full four-year term on December 31, 2024. I was ready to leave the government and continue giving back to my island from a different role—one where I could still make an impact, but with more peace and privacy. I will forever be grateful to my dear friend who made that unexpected call, and to Governor Pierluisi for trusting me with the biggest opportunity of my career. That chapter changed my life, making me stronger, more grounded, and more committed than ever to serving Puerto Rico. **I know Mom and Dad are proud!**

After completing my term, I knew it was time to return to my roots. I rejoined Kimley-Horn in early 2025, stepping into a new role as Strategic Infrastructure Director. It felt like coming home, but with a whole new perspective. I brought with me the lessons of public service, the grit of navigating government, and the vision of what Puerto Rico—and the Caribbean—could become with the right infrastructure investments. Now, I'm helping shape the future of transportation and infrastructure development not just in Puerto Rico, but across the U.S. and the Caribbean.

I'm also still deeply committed to STEM education and mentoring. That part of me hasn't changed. I continue to support young women in engineering and aviation. And remember Monsi from NASA? In 2023, Monsi nominated me, and I was honored to be inducted into the Space Camp Hall of Fame Class of 2024—a recognition that means the world to me because it celebrates not just what I've done, but who I've inspired and those who believe in me!

Through it all, I held onto my faith and the belief that we are called to rise to the moment. This verse carried me through the hardest decisions and the most defining moments of my journey:

"Perhaps this is the moment for which you have been created." — Esther 4:14

To every young professional reading this, especially my fellow Latinas: your story and voice matter. Your dreams are valid. You may come from humble beginnings, but that doesn't define your destination. Work hard. Stay grounded and have self-initiative. And never forget where you come from because that's where our strength lives.

I became the first engineer in my family, and my journey is far from over. There is power in perseverance and managing the unexpected. You never know who you might inspire next.

Connect with Eileen:

The Stone Soup Engineer: A Royal Quest from the 'Lemon' Class to the Cloud

Coauthored by Rachael Orumor Anorue

Technological Innovation Empowerment Catalyst

Hi, I'm Rachael Orumor Anorue, and I Didn't Want to Be Second-Best.

My story starts not with code, but with competition, courage, and color-coded classes.

I had the incredible honour of being raised by parents who championed what I had and what I could become. My early education was primarily in girls-only schools, an environment where I was simply a girl, free from the societal noise that often tells young women what they can't do. I grew up surrounded by encouragement, watching animated movies, rearing tadpoles, and planting vegetables with my mom. I looked at those Disney princesses, not for the princes, but for their courage and ability to do things differently. I wanted to be that kind of remarkable.

When I first arrived at Lagos Anglican Girls Grammar School (LAGGS), my competitive spirit was immediately tested. The school had a distinct color-coding system for classes, designed to help students learn at their best pace:

- Purple was for very intelligent students (The Royalty, basically).
- Lemon was for intelligent students (The Second-Best).
- Then came Orange, White, Green, and Red for everyone else.

After missing the first entrance exam, my scores placed me squarely in JSS 1 Lemon class. I thought, "Lemon? I don't think so." I had to tell myself, "My post-entrance grades were better than some of theirs! Just because I was on the waiting list doesn't mean they are smarter, and I am not staying here!" I truly hated being categorized as second-best. Purple is royal! It was our school uniform color! I tried my best to earn a transfer, but my teacher just said there wasn't much difference between Purple and Lemon. I thought, "Tell that to the Purple-girls that walk shoulder-high during break-time!"

My First Lesson in Leadership and Elegance

I eventually settled into my Lemon class, where I was able to secure the position of the Noise-Maker Captain of the class. My job description was straightforward: write down the names of everyone who made noise in the classroom.

My leadership strategy was anything but strict. I realized that if I enforced the rule by the book, I'd have no friends by the end of the first week. So, I engineered a compromise: "You can discuss, but don't shout. If everyone talks quietly, I wasn't going to write any names." That was my first real-world problem-solving experience, long before I ever wrote a line of code. It wasn't about enforcing a rigid rule; it was about finding the best-fit solution that would carry most people along. It taught me that real leadership isn't just about command and control; it's about consensus and finding creative ways to achieve the goal. It's a principle I use in every boardroom and consulting project to this day.

The privileged training at LAGGS instilled in us an unparalleled culture of meticulous grooming. The principal's constant catchphrase, "**The way you are dressed is how you will be addressed**," underscored the importance of our presentation. No detail was too small: we were expected to have clean white undergarments, perfectly filed nails (with no color, save for transparent gel), sparkling white socks, well-polished brown moccasins, and a clean, sharply ironed uniform. We learned to sit in a way that preserved the pleats of our uniform. The ceremony would conclude with the principal's memorable command: "**Girls… tummy in, chest out, walk like a lady!**" before we marched out chanting our school song, "**Purple and brown is everything I'm wearing...**". This rigorous grooming standard was the crucial starting point of my elegance and style journey, directly informing the polished and intentional disposition I maintain in all my professional and public engagements.

The Royal Veto and the Switch: From Astronaut to Algorithm

When I reached Senior Secondary School (SSS), I finally earned my coveted spot in SSS 1 Purple. I had planted my flag on Mars, and I was going all-in on science. Driven by this motivation, I excelled in my first term of SS1, achieving the highest result across the level with nine A's and one B. This success fuelled my ongoing quest for self-discovery, where I eagerly absorbed knowledge and enjoyed new findings both inside and outside the classroom.

My imagination was massive. I wanted to be an astronaut, intrigued by the discoveries they brought back from space. If that failed, I wanted to be like Michael Faraday and solve Nigeria's persistent electricity problems. And then, my mom stepped in with the most hilarious and logical veto: "I have only 2 kids," she said. "One cannot live in space! And as for electricity, your profession will be anything but dangerous." Her response seemed perfectly fair to me; she has two kids—fair enough! I didn't bother fighting that; I had a million other things I could become, no problem at all!

One subject I genuinely loved was chemistry, thanks entirely to a teacher unlike any other: a brilliant and beautiful woman, who was also a wife and mother. She was my first role model in Science, Technology, Engineering, and Mathematics (STEM), showing me that dedication and brilliance could coexist with a full life. Her patience and guidance solidified my desire to become a chemist. I was so dedicated that I memorized chemical formulas and frequently requested time in the lab during breaks to practice creating chemical compounds. This enthusiasm made me her favorite student. That devotion led to a comical moment when I asked if I could inhale laughing gas to experience its effects. I was already laughing uncontrollably before she even answered. She simply smiled and said, "You haven't inhaled laughing gas, and you are already laughing this much." She gently steered me toward practicing titration instead.

My experience reached its peak in my final year when I was entrusted with major roles: Assistant Head Girl, Yellow House Sports Prefect, President of the Nigerian Red Cross Society, and Secretary of the Student Representative Council. Though I cherished the recognition at the time, I couldn't have known then how profoundly these varied leadership duties were foundational experiences preparing me for the leadership and organizational demands of my academic and professional life.

I left secondary school determined to pursue either chemistry or petrochemical engineering. During the months I spent at home, my mother wisely signed me up for a desktop publishing class and a course in Fortran programming. It was through programming that my true curiosity was ignited. I found myself completely intrigued by the logical world of code and the limitless scope of opportunities it provided.

I soon discovered a strong affinity for programming, though I was initially hesitant to admit that it rivaled my love for chemistry. Eventually, I found a compromise in Computer-Chemistry, enrolling in the pre-degree program at the first university I attended. However, a few months into the session, the Ministry of Education recalled the program and mandated that all students write the Joint Admission Examination.

Unwilling to write the exam, I decided instead to pivot and seek admission for Chemical Engineering at a new institution. During this transition, I sought guidance from the Counseling Office at my prospective university. The counselor who interviewed me asked several questions and then, based on my answers and my Fortran background, suggested that I was ideally suited for programming, recognizing my passion for problem-solving. This unexpected but powerful recommendation settled my path: that conversation was the genesis of my Software Engineering career.

The Defining Moment

The real, life-altering switch happened when I entered the ICT department at the university. After years in the comforting bubble of all-girls schools, I walked into a lecture hall and counted: 40 boys and only 2 girls in my class.

It was a dramatic realization—a cold splash of water on my face—that opened my eyes to a larger society where women were jarringly absent in certain technical settings. In that moment, I discovered the passion that would truly drive my career: working for the inclusion of women in the Information and Communication Technology (ICT) industry.

As I started to immerse myself in the field, I found my inspiration in an old European folklore story I watched as a child: The Stone Soup.

My Stone Soup Philosophy

The story is simple: a hungry traveler got to a starving village. Everyone kept their food to themselves. All the traveler had was an empty pot, which he filled with water, then added a single stone. He kept cooking, and the villagers, driven by curiosity, approached one after the other. He asked them to contribute the vegetables they had so the soup could be sweeter than expected. Everyone gave the little they had, and they shared a feast.

My own journey started with an empty pot, and my single stone was my skills. I decided I wasn't going to sit around complaining about the absence of women in the ICT industry; rather, I was going to start cooking my own "stone soup".

I began by initiating a developers' club at my university. We were only two girls, with anyone else who wanted to join, but it was all I had to start with. I learned new programming concepts and trained others week in, week out until I graduated. I believed if I kept cooking, others would inevitably contribute their carrots, beans, and spices to the pot. This principle became my professional mantra: "Your skills, no matter what, are all you need to **MOVE UP**. It is not about having everything that others need; it is about starting with yours."

My drive remained high throughout university, mirroring my secondary school commitment. Over six years, I simultaneously pursued two rigorous paths: a combined bachelor's and master's program in Software Research and Development Engineering at one institution, while also undertaking a separate degree in Linguistics, specializing in English, French, and Chinese at another. My dad was the driving force behind my Linguistics degree, as he firmly believed that proficiency in multiple languages was essential to position me as a truly global professional. His foresight ensured that my technical skills would be complemented by the global communication tools needed for a limitless career. This intense, six-year effort culminated in successfully earning the degree and a master's in software engineering, alongside the degree in Linguistics.

This unique combination immediately fueled my passion for computational linguistics, which became the genesis of my deep interest in Artificial Intelligence and Deep Learning. As a second-language learner of French and Chinese, I experienced the necessity of automated translation firsthand. In fact, my Master's thesis project focused on developing an expert system to help an English speaker learn the Yoruba Language.

My Work Today: Building the Digital Future of Africa

My professional life is dedicated to ensuring that Africa is not just a consumer of technology, but a builder and a leader. I have over 10 years of experience as a software research and development expert, with a specialty in Artificial Intelligence & Cloud Technologies.

My day-to-day work is incredibly varied; however, it is dedicated to leveraging my deep technical background for real-world impact and sustainable development, particularly within the African tech ecosystem. I specialize in bridging the gap between global technology and African reality:

- **Tech Entrepreneurship**: I am the Founder and CEO of Sens Orbit Company. I was recently honored as the Most Influential Business Transformation Leader 2025 (Nigeria) by Acquisition International's Influential Businesswoman Awards. I focus on guiding clients in adopting software best practices to drive growth and profit for their business.
- **Policy and Training**: I facilitate high-level workshops with organizations like the African Union (AU), UNECA (United Nations Economic Commission for Africa), and Google. I train African policy makers on tailored AI Policy Blueprints and AI Sprinters Frameworks to responsibly unlock Africa's AI potential.
- **International Consulting**: I have consulted with the United Nations Women on training curriculum design for the African Girls Can Code Initiative. I also worked with the Paris-based Organisation Internationale de la Francophonie (OIF) as an Innovation Consultant with mission trips in five African countries and Haiti. I also worked with the Deutsche Gesellschaft für Internationale Zusammenarbeit (GIZ) GmbH as a training consultant, leading a three-day workshop on Agile/ SCRUM Methodology for women tech professionals.

My pursuit of knowledge continues today as I work toward a Doctorate in Information Technology, with a focus on leveraging AI, including robotic systems, to help African farmers—solving real-world, ground-level problems with cutting-edge tech.

Career Highlights and What Was More Challenging

The Highlights (The Sweetness of the Soup)

The "ingredients" that have truly enriched my Stone Soup journey are the moments of systemic change and impact:

- **The Amazon Awakening:** When I won the overall 3rd best student award in my set of 300+ students at the junior high level in LAGGS, my prize was a book: **The Amazons of the New Nigeria.** Seeing a Nigerian woman—an alumni of my school, no less—defined as an "Amazon" was the ultimate real-life proof that I could be one of those courageous princesses I watched in Disney movies. I realized my leadership roles were simply me preparing to become that Amazon. I hoped some girl, somewhere across the globe, would one day feel encouraged because I held a high-profile tech position.
- **Honing Executive Acumen**: In 2022, I solidified my global business acumen when I was selected by the U.S. Department of State as one of 700 young African leaders for the prestigious Mandela Washington Fellowship for Young African Leaders (MWF/YALI). Following intensive Leadership in Business training at the University of Texas at Austin, Texas, I was subsequently selected as one of only 100 Fellows across the U.S. for the Professional Development Experience (PDE). My host company for the PDE was Deloitte in Washington, D.C., where I collaborated with experts to co-create the 2023 managerial and operational strategies for my African tech companies.
- **Pioneering Tech Leadership**: My advocacy began with volunteering. In 2016, I made history by becoming the first woman to lead the Google Developers Group (GDG) in the Benin Republic. This led to me becoming the Country Mentor for the GDG in Benin in December 2017, and the only female

national mentor in sub-Saharan Africa at the time. It wasn't about the title; it was about opening the door for the next woman to walk through. For five consecutive years, I also served as the Ambassador for the Google Women Techmakers (WTM) community in the Benin Republic. This leadership role involved weekly training sessions where I taught new Google technologies. Working closely with dedicated volunteers, we not only educated women in the community but also successfully facilitated their placement in crucial internship and job positions. This position granted me the privilege of attending a Google I/O event in Silicon Valley and speaking at annual GDG/WTM summits across sub-Saharan Africa. Today, I am proud to see a number of prominent emerging female tech leaders who began their journey in those very training sessions.

- **African Robotics Champion**: In 2017, I mentored Team Benin for the FIRST Global Robotics Competition. We didn't just participate; we emerged as the African robotics champion and ranked 7th out of 163 global teams. We used that visibility to successfully raise funds for the Girls Focused Robotics Training Program (GFRTP).
- **Creating Role Models**: Through the femCoders Robotics Initiative (which I founded in 2015) and the GFRTP, we have trained over 2000 students, with 70% being girls. The greatest, most exciting feedback we get from the trainees is seeing young female students being trained by women who look just like them—a female. That feeling is priceless.
- **Global Innovation & Recognition**: I also won the First Prize at the Global Innovation Through Science and Technology (GIST) competition in Dakar, Senegal, in April 2016. This victory validated the power of interdisciplinary solutions and fueled my commitment to local innovation.

More Challenging Than Anticipated (A Little Too Much Salt)

The biggest surprise in my career wasn't the code; it was the constant demand to prove my identity and competence.

I recall vividly a meeting at one of the banks my company consults for. I walked into the meeting room and sat down gently while I waited. A man walked in, looked at me, and asked with an air of certainty,

"Where is your boss?"

I responded, "I am the boss."

He looked surprised and doubled down,

"No, I mean the tech expert."

I met his gaze and replied,

"I am **the** tech expert!"

This kind of interaction highlights the constant, exhausting requirement to establish technical credibility as a woman, especially when leading a tech company in Africa. While this skepticism acts as an unspoken tax on female leadership, it also strongly fuels my determination to pave the way so the next generation avoids facing the same barriers. To challenge that outdated perception, I adopted the online identity, "**GeekOnHeels01**." This moniker declares my presence: that a technical expert can arrive with style and elegance, demonstrating that being a woman in tech is not a bug, but a fundamental feature.

Another persistent challenge was moving beyond training to financial empowerment. My initial Stone Soup fire was "getting low" because I realized training and upskilling alone weren't enough. Many women have brilliant startup ideas, but securing the financial sponsorship for that crucial next step—especially for women's ideas—is a battle I'm still fighting. Consequently, the conversation around diversity and inclusion must mature. It is not enough to simply allocate positions; we must focus on giving **women the financial access and autonomy to create their own ventures and define their success.**

The Feature, Not the Bug

My career trajectory, blending Software Engineering, AI, and Linguistics, has allowed me to contribute to areas I am deeply passionate about: technology innovation and the empowerment of women in STEM across Africa. The recognition I have received over these years has been less about personal achievement and more about affirming that a high-profile geek can indeed be a woman in a position of influence.

It Takes a Village: My Playful Plea for the Next Generation

As I look forward, my work is about creating a truly sustainable pipeline for women in tech. I've focused on systemic change:

- **Empowering the Next Leaders**: By stepping down from my leadership roles at GDG and Women Techmakers in Benin Republic, I ensured the appointment of at least five women into high-level leadership roles in the ecosystem, creating a ripple effect of female leaders across new locations.
- **Training and Mentorship**: I actively mentor young women who have gone on to become successful tech entrepreneurs. I have volunteered as a mentor for young startup beneficiaries of the Tony Elumelu Foundation grant and designed training programs for the UN Women African Girls Can Code initiative, to name a few.

My Final Takeaway

When I look back at my academic and professional learning curve, from the Lemon class to the Cloud consulting, the lesson remains the same, and I share it with every girl I meet: It takes a village, but you have to start the pot.

My mother disapproved of me working in space or solving Nigeria's electricity problem because she saw the danger involved, which steered me toward **desktop publishing and Fortran programming**, and my father's insistence on **multilingualism to position me as a global icon**, together shaped my professional pathway. That seemingly restrictive love ultimately propelled me into ICT and AI—a field that allows me to solve problems just as large (like empowering a continent or increasing crop yields for farmers) without the need to fly a rocket.

The most valuable skill you have is the one you already possess. Don't wait for the perfect moment or the perfect support system. All you need is your skill, your stone, and the willingness to light the fire. So, **light your fire, start cooking**, and the village—the resources, the support, and the success—will inevitably follow.

Connect with Rachael:

From the Mekong to Mentorship: A Journey of Resilience, Curiosity, and Empowering Girls in STEM

Coauthored by Dr. Tracy Nguyen

I. The Mekong River

IMAGINE BEING FIVE years old, torn from everything familiar, unsure if you would ever see your family again. Imagine being separated from your parents in a foreign land, unable to understand why the hands that had always held you suddenly let go. Even now, certain scents carry me back to that moment. The crisp fragrance of an apple instantly returns me to the refugee camps, to the fear, confusion, and uncertainty of not knowing what the next day would bring. The smell of grilled fish on a barbecue takes me there as well, reminding me of the small moments of comfort and community we created amid hardship. These ordinary scents hold extraordinary meaning. They carry the weight of survival, resilience, and hope born in the most fragile of circumstances.

I was five years old during the Fall of Saigon. My father, a high-ranking government official in Cần Thơ, the second-largest province of South Vietnam, had built a life for our family defined by comfort and stability. We were surrounded by nannies, household staff, and quiet privileges that made the world feel predictable and safe.

Almost overnight, everything vanished.

With the collapse of South Vietnam, our safety, our home, and our certainty disappeared. My father refused to abandon his beloved country until Communist forces were at our doorstep. As a senior government official, he knew that staying would mean imprisonment in a reeducation camp, where political indoctrination and forced labor awaited those associated with the former government. When it became clear that remaining would cost him his freedom and endanger our family's future, he made the agonizing decision to flee. My mother moved quietly and decisively, steadying us with a calm she did not feel, guiding their five daughters and newborn son into the night, each of us with a small handbag she had long kept ready. We slipped under the cover of darkness onto a small fishing boat, only three meters long, crowded with 214 people. The boat wound its way through the Mekong River, past Cambodia, and into open waters, every moment shadowed by fear and uncertainty.

At one point, the boat began to leak. Water poured in faster than it could be bailed out. People scooped desperately with their hands, buckets, anything they could find. My mother had only one bar of Dove soap. In a desperate act of ingenuity, she pressed it into the crack. That small bar of soap helped keep us afloat, a tiny, fragile anchor against the river's uncertainty.

Eventually, a U.S Navy ship appeared. We did not know then that it could take only fifteen people. I was one of the lucky few, helped aboard by my father's close friend. My father had already boarded, but when he saw that my mother and siblings were still stranded on the crowded wooden boat, he made an impossible choice. He returned to them, leaving me in the care of his friend. By the time the ship stopped taking refugees, I was standing alone amid the chaos, unable to understand why I was being torn from my parents or why the world suddenly felt vast and foreign. As if that were not enough, Viet Cong propaganda soon spread word that a high-ranking officer, my father, had been killed. I did not understand politics or war, but I understood loss. Panic and grief washed over me as I stood there, my small hand now held by a stranger, surrounded by strangers, the crushing uncertainty of what would come next pressing down on me like the open sea.

What made the moment heavier was what I was holding. In my small hands was a handbag meant for my newborn brother, barely a month old, filled with food and diapers. Unbeknownst to me, my family was heading toward Malaysia, and they would spend seven days and nights adrift at sea before being rescued by the Malaysian Navy.

I was five years old. I did not yet understand fear, uncertainty, or the miracle of survival. I only knew that my family was gone. I was placed in the care of a friend of my father—a man I barely knew—who had suddenly become my only lifeline. We were traveling to Subic Bay in the Philippines, a place others called safe, but to me it felt vast, far away, and frighteningly unknown.

To ensure my survival aboard the rescue ship, my father's friend gave me his last name and listed me as his child. What protected me in that moment also complicated my identity, turning my own name into a barrier to reuniting with my family.

Officers arrived from the French and U.S. embassies, each holding the promise of a new beginning. My father chose the United States, seeking political asylum and freedom. He was set to go to Camp Pendleton in California, carrying hope through fear and uncertainty. Meanwhile, I was already at a refugee camp in Fort Chaffee, Arkansas. When my parents searched the Red Cross camp lists and saw a child under their friend's surname, they trusted their instincts. They changed course from Camp Pendleton to Fort Chaffee. That leap of faith reunited our family and altered the course of our lives.

The moment we finally found each other was indescribable. My parents' faces lit up with radiant smiles, their arms trembling as they held me tightly, as if to make up for every day we had been apart. Tears streamed freely, laughter and sobs mixing into a chaotic blend of relief and joy. Months of fear, uncertainty, and longing dissolved instantly, replaced by the profound, unshakable happiness of being together again, a family restored at last.

I was six when I was finally reunited with my family in the United States. Rebuilding a sense of safety was only the beginning; my siblings and I also had to rediscover who we were. Starting over in a new country was both challenging and deeply humbling. English was our second language, and learning not just a new language but an entirely new culture tested us in ways we could not have imagined. My father, once a respected government official in Vietnam, quietly took on work as a custodian at a Baptist church. My mother went to cosmetology school and became a hairstylist. I did not yet understand the word *sacrifice*, but I felt it in the quiet ways my parents set aside their own lives so that my siblings and I could move forward.

II. Lessons in Resilience

The early years were difficult. I struggled in school and failed kindergarten, carrying the psychological weight of displacement and prolonged separation that shaped how I moved through the world. I barely spoke in class, feeling unsure and invisible in an environment that was entirely new.

My parents worked relentlessly just to keep our lives steady, navigating unfamiliar systems while trying to survive. They rarely had the time or experience to attend school events or extracurricular activities. I am not sure they ever made it to one of my volleyball games, and I never held that against them. They were doing the best they could, pouring everything they had into securing our future. Watching their quiet perseverance taught me endurance early on and instilled a deep sense of responsibility to create opportunities where support and belonging are visible, accessible, and intentional.

What my parents were unable to provide through their physical presence, they gave through conviction. They rarely spoke in detail about careers or long-term plans, but one message was unwavering: education mattered. My father had studied at UCLA before my older sister was born, and in our household, college was never framed as a choice but as an expectation, the natural next step after high school. It was the clearest path they knew toward stability and opportunity, a way to build a future less fragile than the one they had been forced to leave behind. I did not grow up with a roadmap, but I grew up with certainty. Going to college was simply what came next.

In high school, I explored many interests, driven more by curiosity than clarity. I did not yet know who I wanted to become, but I was eager to learn, ask questions, and try new things. By the time I graduated, the only thing I knew for certain was that I wanted to help others through healthcare. That desire did not come from a carefully defined plan, but from lived experience and observation.

As I grew older, my sense of purpose became clearer. I spent my childhood watching my older sister navigate life with severe visual impairment from Peter's Anomaly, a rare genetic corneal disorder that left her legally blind. I've always admired her, not for her disability, but for her determination. She never let her vision define her limits. She pursued higher education at the University of San Diego and went on to earn her PharmD, choosing the harder path because it meant building a life on her own terms.

Seeing her advocate for herself and tackle challenges head-on showed me what healthcare can truly do. It's not just about treating eyes; it's about helping people reclaim independence, confidence, and the freedom to live their lives fully. Her courage and perseverance inspired me to pursue optometry, a field where I can blend science, empathy, and service to make that kind of difference for others.

I took a different path, attending the University of California, San Diego (UCSD) as a commuter student. Commuting allowed me to balance school with family responsibilities and financial realities, and it gave me the chance to support my older sister, driving her and fostering her independence while pursuing my own education. UC San Diego provided a strong academic foundation and reinforced the discipline, independence, and perseverance that had been quietly forming since childhood. I majored in Biology and minored in Music, a combination that reflected both my curiosity about science and my need for creative expression. Music became my anchor, offering balance and an emotional outlet as I navigated the demands of a rigorous science-driven path.

By charting my own path, I also inspired my two younger sisters to follow in my footsteps, each pursuing higher education at UCSD with the confidence and determination I had worked to model.

Although I was effectively a first-generation student navigating higher education without a guide, I was not alone. UC San Diego had a strong Vietnamese student community, many of us children of refugees whose families shared similar stories of displacement, sacrifice, and rebuilding. I found my support system through the Vietnamese Student Association (VSA). Within VSA, mentorship happened organically, confidence was built collectively, and voices that might otherwise have stayed quiet were encouraged to grow.

In many ways, VSA was my first experience with a community like the Society of Women Engineers (SWE). It showed me the power of shared identity, representation, and belonging, how meaningful support can transform uncertainty into confidence. That lesson stayed with me. Years later, when I encountered SWE, the familiarity was unmistakable. The same principles were there: community, mentorship, encouragement, and the belief that when people feel seen and supported, they are far more willing to step forward, lead, and lift others along the way.

III. Parenting and Nurturing Curiosity

Decades later, I realized that these early experiences had planted the seed for how I would parent and mentor. Every gap I had encountered, every moment of uncertainty, became a lesson in the kind of guidance I wanted to provide for my children and for the next generation of girls in STEM.

Over time, I found my voice. I learned to speak up, ask questions, and keep going even when support was scarce. Those early struggles shaped who I am today: resilient, empathetic, and determined. Growing up, my parents worked tirelessly to put food on the table and keep a roof over our heads, so the support I needed was not always there. With my own children, I made a conscious choice to do things differently. I nurtured their passions, encouraged their curiosity, and made sure to be present for every game, competition, and milestone. I wanted to give them the wings to explore, grow, and discover their own paths, something I know my parents would have done if they could. This belief in showing up now guides how I mentor girls in STEM, helping them see that their beginnings do not define their potential.

With that same commitment to nurturing curiosity, I watched my eldest daughter, Madalyn, begin to discover her own passions. When she was nine, she asked Santa for a LEGO Mindstorms robot for Christmas. I made sure Santa got the message. That single gift sparked her love for robotics and engineering and quietly set her on a path that would shape her life. By sixth grade, as her interest deepened, I began searching for role models and a supportive community to guide her. As a non-engineer, I wanted to find women who could inspire and guide her. I wholeheartedly believe that seeing women succeed in engineering is key to nurturing her interest and confidence in pursuing the field.

That search led us to the Society of Women Engineers in September 2015, the inaugural year of SWENext, a program designed to welcome girls under eighteen into the SWE community. SWE is a sisterhood that goes beyond an organization. It nurtures confidence, opens doors to possibility, and provides guidance at the moments that shape our futures. Through SWE, Madalyn discovered role models who inspired her, and I found a way to turn my own experiences into mentorship and advocacy for others.

IV. Mentorship and Leadership

When Madalyn joined her high school robotics team as an eighth grader, she faced challenges immediately. Being one of the few girls on a male-dominated team was not easy. She encountered bias and discrimination and often had to work twice as hard to be taken seriously. Rather than deterring her, these challenges strengthened her resilience and determination, teaching her to advocate for herself and trust her abilities, even when others doubted her.

When she officially started high school in ninth grade, Madalyn and I co-founded the SWENext club at her school. Our goal was to create a space where girls on the robotics team and beyond could explore STEM, find mentorship, and build confidence before doubt ever had a chance to take root. As only the fourth SWENext club in San Diego County, we faced the challenge of navigating uncharted territory, learning through trial and error. Despite these hurdles, the club quickly became a vibrant community where girls connected with role models, developed skills, and saw that they truly belonged in robotics and engineering. The following year, its impact was recognized with the SWENext Best Practices Award for fostering inclusion and mentorship.

Around the same time, we attended our first SWE San Diego event, the 2018 SWE SD Open House, a chance encounter that proved life-changing. Justina Sanchez, a safety engineer attending her first SWE SD event as a member, was impressed that Madalyn's robotics team prioritized safety. She recognized Madalyn's curiosity and potential, offering guidance, encouragement, and a tangible example of what an engineer could achieve. Over the next six months, they collaborated on projects, building mutual respect and trust. When Madalyn expressed interest in an internship at Justina's company, no positions were available, but Justina advocated on her behalf, ultimately opening the door to a pivotal opportunity. Imagine Madalyn, barely a sophomore in high school, stepping into an engineering environment.

That experience not only reinforced her passion for engineering but also changed the trajectory of her journey, showing her firsthand how mentorship can transform potential into possibility.

Experiencing SWE firsthand clarified my own purpose within the organization. I wanted to ensure girls could explore STEM, find role models, and see themselves as engineers before doubt ever took hold. Witnessing Madalyn grow through mentorship reshaped my understanding of leadership. The most meaningful impact is not measured by personal milestones, but by the lives we help shape along the way.

Beyond SWE, I volunteer extensively to support the next generation of STEM leaders. I serve as a Girl Scout Gold Award Advisor, guiding young women to define and achieve impactful projects that benefit their communities. I mentor numerous robotics teams through FIRST, helping students build technical skills, confidence, and teamwork. I also serve on multiple advisory boards, ensuring that students at all stages of their education have access to mentorship and guidance.

Just as my own daughter attained the FIRST Dean's List, the SWENext Global Innovator Award, and the Girl Scouts Gold Award, I have had the privilege of guiding many students to achieve the same milestones. Watching them gain confidence, develop their skills, and realize their potential is deeply rewarding. Each success is a reminder that mentorship, encouragement, and opportunity can transform curiosity into achievement and that the support we provide today shapes the leaders of tomorrow.

With both of my daughters now in college and my household transitioning to an empty nest, I found more time to give back to SWE and the broader STEM community. I started by serving on the Outreach Committee, later becoming Outreach Chair for two years, working to expand access and mentorship opportunities. Driven by a desire to make an even greater impact, I was elected Vice President for two years. I had never imagined or aspired to become President of SWE San Diego. It was only after a nudge from an SWE friend that I decided to run, inspired by Madalyn, who was serving as Worcester Polytechnic Institute (WPI) SWE president at the same time.

Sometimes we do not imagine ourselves taking a step until someone gives us a nudge. That is what SWE did for me. Now I pay it forward.

I give nudges to students, helping them recognize their potential, take risks, and see possibilities they might not have considered on their own. I share with them the confidence, guidance, and support that I received or sometimes had to search for in my own journey.

My journey as a mentor deepened when I met Dr. Mary Isaac in 2019 at WE Local St. Louis, where Madalyn was pitching her Girl Scouts Gold Award Project for the DesignLab Community Engagement Challenge. We met at a coffee station and marveled at the irony that two San Diegans had to travel across the country to meet, especially since Dr. Mary does not even drink coffee. That chance encounter marked the beginning of a partnership and friendship that has been inseparable ever since.

Dr. Mary stepped in during a critical period, offering guidance, encouragement, and steadfast support. She became not only a trusted mentor but also a close friend and outreach partner. Her belief in the work I was doing gave me confidence and clarity during a time when both were hard to find. More importantly, her generosity and leadership showed me what it truly means to lift others up. Inspired by her example, I strive to be the kind of SWE member I once needed.

Today, I have helped launch more than 30 SWENext clubs worldwide and serve as a SWE Counselor to over 20 clubs, supporting students across diverse communities and countries. I also work closely with parents and teachers who want to start SWENext clubs, offering the practical guidance, reassurance, and mentorship I did not have in my early years. This work is deeply personal to me. My goal is simple: to ensure no student has to navigate this journey alone and that every child feels seen, supported, and empowered.

After high school, Madalyn began a new chapter at Worcester Polytechnic Institute on a full merit scholarship. Almost immediately, she faced challenges that echoed her earlier experiences. Some male peers questioned whether she truly earned the FIRST Community Scholar recognition, attributing her accomplishments to her gender or ethnicity rather than her talent and hard work. For the first time, these doubts triggered imposter syndrome, making her question her place in a space she had fought so hard to enter. But instead of retreating, Madalyn leaned into WPI SWE, discovering a vibrant sisterhood of over sixty peers who shared her passion, welcomed her, guided her, and championed her potential. Through this community, she regained her confidence, transformed adversity into opportunity, built resilience, and sharpened her focus, laying the foundation for both personal and professional growth.

From her freshman year, Madalyn volunteered extensively, honing leadership skills and deepening her commitment to service. Her journey culminated in serving as president of WPI SWE, a testament to how mentorship and community can turn doubt into empowerment. Watching Madalyn lead inspired me to step forward in my own capacity, reminding me of the importance of showing the same support, encouragement, and belief in others that I hope they experience.

V. Full Circle

As I reflect on my journey, from a frightened five-year-old on the Mekong River to a mentor, Diversity Champion, and leader in STEM, I am filled with gratitude and resolve. Every challenge, moment of uncertainty, and act of guidance has shaped who I am today. My story stands as a testament to the power of community, mentorship, and resilience, and to the belief that when we invest in others, we create lasting change.

My greatest accomplishment is not the awards I have received, including the SWE Engage Advocate Award, the SWE Spark Award, the SWE SD Diversity Champion, the NCWIT Educator Award, or others. While meaningful, these honors are not my measure of success. My true measure is seeing my daughters thrive in college, with Madalyn about to graduate and join SWE as a professional, and watching my mentees flourish at institutions such as MIT, Stanford, UC Berkeley, UCLA, UC San Diego, and beyond. Their growth, confidence, and accomplishments are the clearest evidence of the transformative power of mentorship and the ripple effect it creates when they, in turn, lift others and build inclusive communities.

I have always believed that parents make a profound impact. We often make quiet sacrifices to give our children opportunities we never had, nurture their passions, and help them find their voice. Though

I am not an engineer by profession, I am a woman in STEM as an optometrist. Every day, I approach challenges with problem-solving, critical thinking, and creativity, which are the skills I encourage in young engineers: persistence, observation, innovation, and resilience. In this way, my work in healthcare mirrors the mindset I hope to instill in the next generation.

As we celebrate ten years of SWENext, my journey has come full circle. What began with a LEGO robot and a search for role models has grown into a life shaped by opportunity, purpose, and service. This year, after a decade with SWE, Madalyn graduated from Worcester Polytechnic Institute, stepping into the next chapter as both a professional and a mentor. Her path reflects the values that have guided our journey from the very beginning: curiosity, perseverance, and a commitment to lifting others as she rises. Her recognition by SWE, including the Rising Collegiate Star Award in 2024 and the Outstanding Collegiate Member Award in 2025, demonstrates the transformative power of early encouragement, belonging, and community. Seeing my first mentee grow into a confident leader reminds me that mentorship does more than guide; it multiplies, inspiring new leaders and creating impact far beyond a single moment.

At the same time, while Madalyn served as President of WPI SWE, I had the privilege of serving as President of SWE San Diego. That shared leadership felt deeply affirming, reflecting the values we both carry: community, mentorship, and advocacy. In my role, I work to ensure every member, professional or student, feels seen, supported, and empowered. Collaborating with partners such as the National Society of Black Engineers, Society of Hispanic Professional Engineers, LatinaGeeks, Latinas in Tech, and others, we create opportunities, amplify diverse voices, and make inclusion tangible. What I once provided for my children, support, guidance, encouragement, and space to soar, I now extend to young engineers navigating their own paths, helping them see what is possible, avoid pitfalls, and cultivate both confidence and resilience.

I believe the best leaders are the ones who lift others as they climb. In SWE, SWENext, and our community work, mentorship flows naturally, connections come before credentials, and diverse voices rise.

When people feel they belong, they step forward boldly, take risks, and become the leaders the world needs.

The work is far from over. For every girl who hesitates to raise her hand, for every student unsure if she belongs, there is an opportunity to step in, to mentor, champion, and inspire. Every day, I carry with me the lessons of my past, the courage of my parents, and the spark of curiosity I nurtured in my daughter. Even in the face of uncertainty, love, guidance, and persistence can light a path forward.

From the Mekong River to classrooms, labs, and professional networks, my journey has truly come full circle. I embrace my roots as a refugee and immigrant, never forgetting where I come from. What was once given to me—guidance, encouragement, and opportunity—I now pay forward. Each act of mentorship, each open door, and each moment of support ripples far beyond the individual, shaping futures, communities, and the world of STEM itself.

The Mekong taught me perseverance, my parents taught me generosity, and my mentees and daughters teach me every day that leadership multiplies when shared. Every conversation, every word of encouragement, every space where someone feels seen and valued matters. It is in these small but powerful acts that the seeds of confidence, curiosity, and resilience are planted.

This is not just my story; it is proof of what happens when we lift others as we rise. It continues with every girl who dares to raise her hand, every young engineer who finds her voice, and every mentor who chooses to show the way. Together, these moments spark a legacy of empowerment, a chain of possibilities that stretches across generations and changes the world in ways we can only begin to imagine.

Connect with Dr. Tracy:

Raised by Circumstances: How Ganas, Discipline, and Choice Led Me to Freedom

Coauthored by Daisy Galeana

Senior Program Manager

I did not begin life with choices but instead with unique circumstances. Circumstances raised me.

Some of my earliest memories are not of comfort, but of endurance. It was sometimes very hard to sleep when I was a young child. For one thing, my bed was nothing like a mattress. It was instead a straw mat on an unforgiving dirt floor. I would often find myself lying awake and staring at a wall six feet away. For years, the only light my parents could afford in our family's single-bedroom house was basically a piece of camping gear: a candle fueled by a small tank of kerosene. The candle's flame flickered unsteadily all through the long, quiet nights. It cast shadows that danced across the wall. On a particularly sleepless night, I could watch those shadows shrink and grow and stretch and collapse for hours. As a five-year-old, I could imagine the flickering shadows as living things. Their shifting shapes made me think of my future: something undefined but persistent, something pointing forward but still at the starting line. I could not yet describe what I felt, but I could feel it deeply. It was a quiet and curious kind of passion, an inner pull toward something I couldn't yet name.

We lived in a poor village on the banks of a large river. There happened to be a makeshift runway on the outskirts of the village. Fenced in and protected by armed guards, it belonged to a rich family that lived a few miles away. As a little girl, I would watch with curiosity as private planes taxied and took off from that simple runway. I was fascinated by the different small aircraft, observing what their different engines looked like, how they took off, how they changed directions, and how they landed. No one I knew had ever flown on a plane. These large complex machines intrigued me just like the kerosene-powered candle and the small flickering shadows it cast all night long against our silent bedroom's walls. Curiosity, however, did not have a welcome home in my parents' house. Life was about survival and necessity rather than dreams and fascination. We were simply trying to survive one day at a time. The persistent questions of a curious kid often received no answers. As a result, my dreams had no audience. My unanswered curiosity and bids for validation necessarily turned inward. I became the kind of observant child who listened closely not only to what was said, but also to what was left unsaid. And when a lot of important things went unsaid, I learned to read between the lines and navigate the uncomfortable silences.

What might have been a child's temporary coping strategy turned into something both essential and enduring. These early experiences forced me to build an internal reservoir of strength, the kind of self-validating strength that does not depend on constant outside praise, permission, or approval. That strength became my refuge. My *ganas*!

Eventually, our family's precarious living conditions led my parents to place me in a Catholic orphanage. Life and meals were secured, and I finally got to sleep atop soft bedding. The nuns who ran the orphanage were certainly exacting, but they weren't cruel. Following their rigid discipline simply became instinctual. Even silence had rules attached to it, and here as elsewhere, the nuns' rules and expectations were non-negotiable. Everything ran on a precise schedule. Free time was rare, including time to daydream or to just follow one's imagination and curiosity wherever it might go. In that orphanage, as at home, dreams were not discussed. A different future--hopefully a better one--could only hover somewhere off in the distance beyond the visible horizon. That said, the nuns taught me many important things, including how to endure without complaint, how to show up prepared regardless of how I was feeling, and how to finish what I started. My steady development of academic skills and personal discipline was quietly reinforcing a backbone of personal power. It would need to be strong enough to bear a completely unpredictable future in a life beyond the orphanage's secure walls.

When I arrived in the United States at the tender age of 14, the relevance of everything familiar seemed to disappear all at once. The only language I knew was no longer enough; the customs I had learned were now the exotic ways of a foreigner; the unspoken rules of my old family life and my years in the orphanage were all in need of complete revision. On my first day of ninth grade, the vice principal called me and my mother into her office. After letting my mother answer some questions about my life and my prior education, the American woman turned to me. She had more questions. At first, I was silent. She then looked directly at me and said words I would not understand until years later: "This is your life. This is your education. You need to use your voice to make sure your education is a high priority."

I nodded politely. I had been raised to listen politely to authority figures and not advocate for myself. But the vice principal's words stayed with me. Her encouragement and her challenge planted a seed. My education belonged to me. It would become my responsibility, and I could not fulfill this obligation through silence. I would need to speak up in class and elsewhere. I would need to learn to advocate for myself. The seed took time to grow.

The turning point in my life did not come dramatically. It arrived through people like the vice principal and marvelous middle school English teachers who offered belief instead of limitation. They treated my future as an inspiring and open horizon rather than a fixed target. Discipline without vision taught me how to endure. Discipline, paired with the opportunity to excel at school, taught me how to build. Education became the place where everything converged. Learning English as a second language tested my confidence daily, but my *ganas* carried me through it all.

When I began my bachelor's degree in mechanical engineering at San Diego State University (SDSU), I knew I was entering a world not designed for someone like me. My freshman cohort in mechanical engineering had 250 students and exactly one Latina—me. The university is in a county that's about one-third Hispanic. I was a first-generation college student, younger than many of my peers, and I spoke English with an accent. To put it mildly, my personal and family history was quite different from that of the typical engineering undergrad. And yet I knew that I belonged.

MESA Mathematics, Engineering, Science Achievement reached out to me and became a north star of hope at SDSU. The program's wonderful people guided me, advised me, provided resources, and built a community around me. That academic community became my new village, a village that encouraged curiosity and daydreaming. This new village did not change who I was; it strengthened who I was becoming.

While working on my bachelor's degree in mechanical engineering, I faced a harder personal challenge than any of my classes. My role in a highly traditional Mexican family demanded my undivided commitment toward my family, including babysitting on an ongoing basis, contributing to the family economy, and essentially standing in for my parents as the oldest child in a growing family.

Laboring to meet those extraordinary family demands would have cost me my education. In response to my decision to stay in school and finish my degree, I was kicked out of the house. I was made to feel guilty for not "maintaining tradition." I felt like an outcast. Once again, my parents failed to see a different future for me from what they had known. As I saw things, leaving home to stay in school was not an act of rebellion but one of survival.

After graduation, to follow another old tradition, I quickly became engaged. This time, it was my American girlfriends who helped me realize that I didn't "have to get married just because I turned 21." I had the power to choose my own future, not as a woman or a Mexican, but as an independent human being. Little did I know that turning down this opportunity would end up opening a new door. Applying to and then attending graduate school became my first fully conscious act of authorship.

After graduating from college, I moved away and started a full-time job as a test engineer in Florida. I also began classes en route to a master's degree in engineering at the University of Miami. A stark challenge arose when my manager refused to grant me a more flexible schedule. I persisted at my job and took classes during my lunch hour and before or after work. I managed to complete the master's degree in two and a half years while working full-time.

My Ph.D. journey was also full of challenges. I couldn't afford not to work while studying for the degree, so I continued working as a test engineer while also being a full-time student. At one University of California campus, my graduate advisor demanded I resign from my full-time job to dedicate myself exclusively to my doctoral work. It was what all the other doctoral students were doing. It was not a formal requirement, however, and I couldn't afford not to work. I continued working full-time while conducting all my research, and all the while being a divorced single mom with a mortgage and two children at home. The graduate stipend alone could not cover my fixed costs and, unlike grad students from affluent families (of which there are many),

I couldn't turn to my parents or other family members for financial assistance.

I recalled the moment I refused to resign from my full-time job. My academic advisor responded with anger. He wanted me gone and described my leaving the doctoral program as a simple and inevitable act of "natural selection." I disagreed and called it a simple act of discrimination. Despite the setback, I ended up happily transferring to a joint doctoral program between UC San Diego and San Diego State University. By the time I completed my doctoral work in aerospace and mechanical engineering, I fully recognized that discipline and effort had become my most important choices. Not only my career but much else besides unfolded as the continuation of a series of choices. Discipline and effort revealed a new road to freedom.

Parallel to my academic development, I began working at Solar Turbines Inc. First, I became an engineering intern and then a full-time professional team member. Theory met reality in this job. Accountability and precision mattered, and discipline was essential. My new position in the oil and gas industry was significant because I was now competing in the major leagues. To survive and thrive in this job, I would have to leave behind all the traditional labels that would inhibit the progress of someone like me--labels like female, Chicana, and mother. My identity shifted from who I was expected to be to what I could deliver. Instead, I would have to prove myself through unimpeachable competence, accountability, and results. At Solar Turbines Inc. I had the opportunity to complete a rotation program in development test, design, combustion, materials, and instrumentation pertaining to new product development. It was fun to finally arrive and be able to add to the team. I didn't know the entire gas turbine engine yet, but I was willing to learn--and learn quickly. I remember conducting failure analysis in the lab for a combustion liner using scanning electron microscopes and finding the failure. Presenting the results gave me a sense of accomplishment. My next task was to help re-design and test new fuel injectors and analyze the test data. This project gave me the opportunity to learn new software and programming like LabVIEW. I was being paid to undertake complex engineering projects that also struck me as fun.

I eventually moved from that position to B/E Aerospace, where I worked as a test engineer conducting flammability, static, and dynamic testing for private jets. In this kind of work, nothing is more important than guaranteeing safety. This job was granted to me because of my bilingual skills. In Miami, the workforce is predominantly Cuban, and they speak Spanish. I was hired to be able to speak with the hourly employees and do my job as a test engineer, which was fantastic. I had the best of both worlds, speaking my native language and making progress in the world of private jets. This job was more hands-on. Conducting tests under the supervision of the Federal Aviation Administration (FAA) gave me the opportunity to lead new ways of doing static and dynamic tests. I was challenged to write great test plans that the FAA would be able to approve on my first try. I worked hand in hand with the FAA representatives, who were eager to teach me about the process. After a year and a half of working at B/E Aerospace, I got my first memorandum of understanding (MOU) with the FAA and my own FAA project number. I was on my way to a great future as a test engineer.

I then joined L'Garde, Inc., as the program manager for something called a Solar Sail. As the name suggests, this is a spacecraft propelled by sunlight. In December 2024, NASA launched the Solar Sail I had overseen building. I oversaw manufacturing the prototype, documenting every step of that process, and then leading the tests on the prototype. The manufacturing process was a challenge because the material was Mylar and the thickness was 20 microns, which is 20 times thinner than a human hair. I spent months at NASA's Glenn Research Center making sure every detail of the prototype was ready for the test. I prepared by overseeing the vibration test, ascent vent, thermal, and vacuum tests. As you may imagine, a project like this only gets one chance to deploy, since when it does go to space, it has to work. There are no second chances. I remember standing inside NASA's vacuum chamber (the largest vacuum chamber in the world) and feeling the full weight of being the test leader. It made me feel like a giant--albeit a giant who hadn't slept for days. Nevertheless, I was confident the prototype would work as planned.

The child who had once imagined shadows dancing on the wall

from the light of a candle was now helping to put a spacecraft into space. Learning that the Solar Sail had indeed been launched into outer space was far more than a cool technical achievement for me. I broke through a glass ceiling. As a Latina engineer who had grown up in a cabin with a dirt floor and no electricity, I felt I had traveled an extraordinary distance just as that spacecraft did. I, too, had finally achieved lift-off. Right there, the bigotry and low expectations of others dissipated into thin air.

When I returned to Chula Vista to be closer to my siblings, I found a job as a senior engineer with BFGoodrich Aerostructures in the Airworthiness Certification department. The projects were easier for me since I already had similar experience under my belt. I felt comfortable working with the FAA and following the Code of Federal Regulations (CFR) closely. I led Boeing 787 and Airbus A380 thrust reversers and nozzle tests. Now I was the expert, witnessing tests and reviewing data. My level of expertise on the difference between the FAA and the European Union Aviation Safety Agency (EASA) was the highest. I was responsible for ensuring that designs met the highest safety standards before flight.

Later, I returned to Solar Turbines, moving from my earlier position of test engineer into a management position. Like starting out in engineering as a college freshman, I was once again breaking into a boy's club and succeeding through skill, persistence, and discipline. I was accountable for test systems commissioning. This time, I had the chance to learn everything about the entire gas turbine engine. Thirsty for knowledge, I added more software and programming skills to my toolbox. I spent most of that time in the test cells, learning about engine tests and how software can be automated to collect data in real time. I also used my teaching skills and trained over three hundred technicians and engineers on how to use the latest version of test data acquisition.

I became a principal development test engineer. In that position, I coordinated engine assemblies, organized test facilities, and conducted engine tests while working closely with the combustion, material, compressor, turbine, and design departments. I was leading fifty-four test cells and tested over one hundred engines per year. New software kept appearing that allowed us to test faster and collect quality data live. By this point, I had serious experience and a clearer sense of the direction I could steer my career in through discipline and determination. I wrote and then followed a road map toward upper management.

I was promoted as the regional technical manager for Mexico and Latin America regions. Once again, I was able to use my bilingual skills and learned Portuguese to communicate better with my Brazilian team. In this position, I led a team of field engineers and managed accurate forecasts of revenue and department budgets. I interfaced with customers to maintain strong relationships while executing contracts in a timely manner. I was responsible for adherence to contract service, operations, and performance specifications, as well as technical metrics specific to the contract.

While climbing the career ladder as an intern, a junior, and then a senior engineer, and then a manager of engineers, I noticed something disquieting: I was alone. Where were the other women who looked like me and had a story even remotely like mine? I paused. Reflecting upon my own story, I realized how complex and devastating the answer was: young women in our communities were never told they could amount to anything, let alone become engineers. So, I rolled up my sleeves and began a new kind of engineering project.

With assistance from MESA, I created Solar's Young Women Academy (SYWA). What began as a small idea about recruiting employees from under-represented communities quickly took flight and became a serious and enduring project. SYWA was a forum through which minority female high school students could rededicate themselves to their formal education while receiving the support, mentorship, and resources necessary to lead them toward solid and well-paying employment in a STEM field. MESA and I designed, planned, and executed an annual program with the sponsorship of Solar Turbines Inc. SYWA was more than just exposure to the world of engineering. It was preparation for joining that world. The academy showed its high school participants that engineering was not a distant concept reserved for others but a viable future for them.

The program for under-represented students was later expanded and replicated throughout San Diego County. It had an impact far beyond what I had initially dreamed. For example, SYWA inspired Caterpillar to create their own version of the program. Other Fortune 500 companies, globally, soon began developing similar initiatives. What started as a response to absence became a movement rooted in visibility and access. In that moment, the meaning of leadership changed for me. Leadership was no longer only about my own personal career advancement up the corporate ranks. It became more about working to change the systems so others would have the same opportunities I had created for myself without having to face all the same battles alone.

My path then took me to Rolls-Royce in Indianapolis. Working as Head of Planning, Assembly, and Test, my responsibilities included shaping systems and strategy and building teams to bring these plans to fruition. I was the focal point between business programs and overall execution plans, including overseeing the load and capacity for on-time delivery, and applying process improvement techniques and driving business improvements.

At present, I work as a Program Manager for GE Aerospace (General Electric) in the field of New Technology Introduction. The program's mission is to help guide advanced technologies from concept to reality. The position involves leading the integration of contract deliverables to support assembly and operations by implementing the strategic direction and organizational capability of test cells with a strong safety culture.

When I reflect on my decades of experience and the variety of job positions and titles I've held, I can clearly define my career trajectory as a record of decisions. I had to decide to pare back feelings of guilt in choosing education over tradition. I had to decide to choose authorship and independence over following the lesser expectations of others. I had to decide to play in the corporate big leagues, learn how to fit in and succeed with integrity, and then break every glass ceiling I encountered. I decided to help other girls and women discover STEM and made my work in that area part of my definition of success. I had to decide to choose my heartfelt "*ganas*" over a simple "I wish."

I often think back to the child who stared at dancing shadows on a wall by kerosene light and who admired the small planes she watched take off from a makeshift country runway. That girl did not really know the word *choice*. She only knew *ganas*; that deeply felt drive to imagine something more when nothing around her could warrant those dreams. That young girl's *ganas* kept propelling me toward achieving not only small dreams but larger ones too. Discipline and determination allowed me to accomplish lift-off and reach far beyond the usual expectations and boundaries.

Connect with Daisy:

A Life Engineered for Impact

Coauthored by Dr. Mary Isaac, PhD

My first memory of the magic known as electricity was coming down the stairs and seeing swirling colors reflecting off our sparkly silver artificial tree very early on a snowy Christmas morning in 1957 in upstate New York. We crept down the stairs, hoping to see Santa, as quietly as two kids under the age of four could be, and it was BEAUTIFUL, although Santa was gone!

What I remember most vividly, besides the tree, were the cardboard blocks colored red and black to look like bricks my parents assembled that we could build things with, and the Play-Doh I would use to stick them together so they would not fall so easily. With walls three or four bricks high and a sheet over the top, it was a pretty good hiding place from crawling babies, which my mom had almost yearly for the next seven, for a total of six girls and two boys.

Thanks to those crawling, crying, ever-needing-a-diaper-change siblings, dolls did not interest me as much as the place's dolls got to live in, specifically dollhouses. And climbing trees with a book where the littles could not go. Once we got old enough, we practiced architectural engineering by building forts that did not collapse in 3-5' snowbanks that the plows would leave in front of our driveways in our lower middle class neighborhood, and watching the house next door being built, then climbing through its different levels at night when the workers weren't there to guess where different things were.

The 1964 World's Fair happened the year I turned nine amid the Cold War, and being so close to Flushing Meadows in upstate New York, we spent two days going through the various technology exhibits, including Disney's It's A Small World with its fascinating mechatronics that would wind up in Epcot Center, and the audiovisual and telecommunications advances being made by companies like AT&T Bell and RCA. In retrospect, my dad, being an electrical engineer, probably had something to do with us going for two days so he could take it all in, and we were just along for the ride.

My interest in how things worked was sparked by that, and the books my grandmother used to buy us that explained how things worked, but no more and maybe less than my growing interest in music, social justice, and protecting our less-able-bodied people and the environment. Then I learned to play guitar for folk mass, rock and roll, the civil rights movement, high school, and 'women's lib', followed by a year of community college. I had scholarships, but my parents did not make enough money to be able to send me and made too much money for the scholarships to cover the cost at Syracuse University, which is one of the few schools that offered biomedical engineering at the time. Instead, I attended a nearby community college for a year, met the man I would marry three years later, dropped out of college, and then took off to explore the world with my guitar.

In late 1975, after running out of money playing music on the road (and calling my grandmother for bus fare home), I was a live-in home health aide until hearing from my dad about the local turbine generator factory in town looking for women and minorities as apprentice machinists and draftsmen. Fortunately, my math grades in high school were good, and I had taken drafting for two years, so I knew how to read drawings. Over the next three-plus years, I began learning the power generation business from the inside out, from the turbine to the generator, not just the hardware, but also the backroom processes that delivered a complete machine for a customer.

After work, we attended classes at a local high school in metallurgy, machining processes, and even public speaking. Several years later, I learned that most of the women had been funneled into the high school route (vs. community college) to allow the men to have the associate degree that went with it. At 21, engaged to the man I met three years earlier, fighting double standards was not yet high on my list of priorities. Within the next two years, those priorities would shift yet again when I had my first and only child. Life was oh-so good until it wasn't.

As a single person, the only person I had to worry much about was me; once married, additional considerations had been added to important life decisions. Sometimes, I felt like I had traded in a piece of my identity to become a "we", so when I got pregnant, I already felt a little fragmented and conflicted. While pregnant, I struggled with losing control of my body to the developing person inside me, but once he was here, the nurturing side surfaced that I was worried might never show up. However, when I returned to work as a nursing mom following the birth of my son the year before, there was no place to store lactation equipment or refrigerate the milk after I pumped, so I gave it up by the time he was six months old. Childcare options were nonexistent, but with a supportive mom and mother-in-law to help, life mostly hummed along until March 1979.

Does anyone in the US remember waiting in line for $10 of gasoline on only odd days during the several gas crises in the 1970s? How about the Three Mile Island nuclear incident? In March 1979, my son was not quite one year old. In the weeks following the Three Mile Island nuclear plant accident and the overthrow of the Iranian monarchy by the theocratic Islamic Republic, the worldwide energy crisis reached a critical point, as did I in my journey, one of several over the next several decades.

Sales of power generation equipment had slowed sufficiently such that apprentice machinists could not even be assured of a third-shift sweeper position once they completed the program. The man I had married three years earlier was a good father, but I no longer wished to be married, nor dependent on someone else to take care of me and mine. It would not be long before I discovered that some contracts may take longer to dissolve or supersede than others, especially when children are involved, but that having a contract was very important in both business and personal matters.

While the training I had received, and the skills I had gained, would stay with me for my lifetime, there were many questions in my mind about what I needed to know to be able to insulate myself and my son from the bad juju in the world, along with being conflicted about how to get there: music or engineering. It's hard to say which interest came first to me as a kid, then a young adult: making music or building things. Having spent a year or so exploring the world of semi-professional musician-as-performer before getting married, knowing what it would take to make a living at it and not wanting my son to have to make any sacrifice, the compromise was practicing my music as a hobby and studying engineering to learn how to start over if needed, by building things that moved, like pumps, wheels, and all sorts of power generation and conversion machines.

Fortunately, the world had settled down to some extent by the time my graduation rolled around a few years later and was maybe even recovering, because an opportunity to work construction at a nuclear power plant surfaced in the spring of my senior year, one that was less concerned about my "B for Balance" GPA and single motherhood than impressed about my ability to read prints and specifications. As a fledgling nuclear field engineer, the training consisted of three months learning the basics of boiling water reactor design and maintenance, after which Painesville, Ohio, became our home during construction until the first unit was ready for turnover to the customer.

Once again, childcare moved to the forefront as a priority; money was not so much an issue as finding available services. Remember, this is the early 1980s, no Internet yet? Until he was ready for school, a nearby church program provided the needed care for a few hundred a month; after that, a mom in the neighborhood picked him up with her son, and he sometimes stayed over if I was on night shift toward the end of that tour. However, it was tough not having family around to help, so when another opportunity came up for me to move to a service shop training assignment back in my hometown that involved turbines and generator maintenance, something I was familiar with and much closer to family, I jumped at it, moving back to NY temporarily for two years, then out to Cincinnati as a turbine engineer when I finished the training program.

Around the same time, my divorce from my son's dad was amicably finalized after seven years of legal separation with joint custody and me having primary domicile, and we both remarried within a year. After a two-year stint in the Cincinnati area, we headed back to upstate New York for what would become our home for the next 15 years, as different roles surfaced for me that piqued my interest and leveraged my talents. There was a brief two-year residence interruption for a sales assignment in Minnesota and North Dakota, where I played a lot of customer golf in the very brief period known as summer. My husband was typically on assignment ten months out of the year and might be home for a few weeks at a time every few months when he was working, but we worked out a system that worked for us. When we had fun as a work or personal family, we had a lot of fun!

That is not to say there were no bumps along the way, but most of the time I felt valued at whatever role I was in, whether it was securing the first LM6000 power plant order in Minnesota and getting to take clients to the World Series, Super Bowl, Final Four, and PGA Masters over twelve months; designing and rolling out a Customer Management System, followed by a Risk Management tool, to help the business forge its way through the power bubble of the late 90s; completing a memorandum of understanding for the second phase of an advanced gas turbine power project in India; or running a $2 billion service sales organization.

Yep, it was fun 'most of the time'. When it wasn't, crying in the ladies' room was always an option, but occasionally, that was not enough. Like the time the production clerk took my turbine out of a machine queue for someone else's. The air turned blue when I found out, and got even more colorful when he huffed, "I won't have a lady talk to me that way!" And it wasn't much fun when someone I had trained got a role that I coveted and was well-suited for. That one ended up with me volunteering for a job elimination in the next year when the opportunity arose. After six months of contemplating my options with a son getting ready to head to college, the person I had trained went to the Big Guys to bring me back, got me a SIGNIFICANT raise, and more flexibility in my role, helping build and service power plants in India. He is a friend even today.

A few years later, as we were about to sign a deal adding capacity and a long-term service agreement to an existing contract, we received unsolicited, incontrovertible proof that at least SOME salespeople had brains when a magnetic resonance imaging (MRI) scan revealed a golf-ball-sized tumor behind my right eye, which had been swollen and tearing. Although not painful, it distended my optic nerve and muscle, caused the eyeball to press against the tear glands, reduced my peripheral vision, and appeared to be growing through the skull, so it was removed and determined to be benign. Three weeks after surgery, the whole sales team met in London to sign the contract for the deal in India, with me just happy to be there.

I must have gotten a few courage points for that, or maybe my resilience was peaking, because the next five years were exactly what I needed career-wise. My son was finishing up college, and my star was rising in my mid-forties, when my husband moved from being a road warrior in the nuclear service business, traveling a lot to Japan and other operating nuclear plants, to take an assignment in Memphis, close to his kids and grandkids. We found a place near Memphis and planned to retire there when the time came (he was seven years older than me), but in the meantime, my career trajectory had me stepping into leading the development of a method of tracking projects through their life cycle from inquiry to enable improved decision-making about which orders to take and which to reject, which included joining an exciting acquisition effort when we bought a company in France, so it took a few more years until I followed him to Memphis.

What finally took me there was my next (and favorite) role as the general manager for service sales in 19 US states, with offices in Houston and Chicago, and home in the middle. The tragedy of 9/11 occurred during that role and transformed this mechanical engineer into a road warrior-cum-leader of a solid team of diverse people who respected each other. Although still only one of a very few women in executive-level roles in our business, more faces looked like me and other underrepresented groups as customers and in leadership meetings each year, thanks to the efforts of the company's affinity groups, which had been instituted and supported by the long-time company CEO.

Following his retirement at the turn of the century, though, the business tenor changed with reports of class-action suits from one, then two of the underrepresented groups. The suits were settled to the satisfaction of the company, but maybe less so the disgruntled employees who raised them. One positive outcome was that some effort was made to introduce more transparency into the promotional process that took leaders into senior executive ranks, so that it became less about who someone knew and more about how they compared to others interested in the same role. Even so, everyone knew that asking a company that had been doing things one way for over 100 years to change how they did it would take some time to be strategized and thoroughly implemented, if ever.

By the middle of the first decade, my boss had been promoted, and his successor turned out to be a bully whose rude behavior deteriorated further when under the influence of alcohol; we might consume a bit of alcohol when we all got together at least once a year, so we were exposed more than once. Several of us began exploring other roles, some better than others at being covert. After a particularly grueling call where he berated one of my peers, when my Human Resources contact probed why I wanted to move, given that she knew it was my dream role, "It's a hostile work environment!" tumbled out of my mouth before I could stop it. They did an investigation, agreed that he needed "rehabilitation," and they put him on a brief plan, then promoted him to vice president. That was a turning point for me with respect to my career and my attitude regarding corporate lack of scruples.

After a somewhat humiliating maneuver where, instead of allowing me to simply transfer to a different role closer to my husband (who had moved us to Wilmington, North Carolina in 2005), I was forced to re-apply for a General Manager role I neither wanted nor was selected for when our five US regions were consolidated, and then other opportunities opened up. The whole process was a dark time for me and forever changed my attitude about the company, but having a strong network enabled me to land softly as the company's executive hand-holder for a large utility customer potentially interested in placing a large order for new equipment and services.

Unfortunately for the company, over the next few years, the customer expected reliability and good service, which the account manager and I could not deliver alone, and when another supplier came along and promised to deliver that at a lower price, it was a no-brainer for the customer to leave us at the altar. My customer was quickly dropped as a critical account, my role was eliminated, and the account manager was re-assigned. When HR contacted me to let me know my role had been eliminated, but that they had me teed up for a few others located somewhere else, the thought of having to become a commuter from a small coastal airport drove me to jokingly ask if I was old enough yet to retire.

Unbeknownst to most employees, the company union contract (which loosely governed how they handled white-collar HR issues) then had a 'golden rule' policy that, if age plus service was 80 or more, anyone could retire and was entitled to full benefits. They would be eligible for whatever benefits they were receiving at that point, such as insurance at employee rates, incentive compensation, and investments, until such a time as agreed, at which point they had to begin active retirement if they were not already past the normal retirement age. HR got back to me within a few weeks to let me know that by the time everything got processed, I would fall under the golden rule parachute, and they could not stop me if I wanted to retire.

At age 52, with 28 years of service and everything I had worked for mostly intact.

By mostly intact, I mean physically, fiscally, and intellectually, but emotionally, things were a little broken. Several more life decisions got made before and during this time, though. From the middle of the first decade of the century until we parted in the second was the longest time my husband and I had been in proximity in years, and we had both changed. Adding to the stress was that we had purchased what we thought would be our "new" forever home at a market high, then had the bottom drop out of the market, and developed an upside-down mortgage around the time I separated from the company. Oh joy. My package had benefits that would allow us to hold on for a while before I had to seriously consider working again.

It's hard to remember when we agreed that he would move out so that I would not have to directly observe as his health deteriorated. We grew apart as often happens, but we did not wish each other harm. He continued to work, while I entertained folks at nursing homes a few times a month, rented rooms to moms trying to start over, trained therapy and service dogs, and thought about what I wanted to do for the rest of my life. One thing I knew for sure was that I did not want to work for The Man, whoever. The other thing I knew for certain was that it would involve girls, STEM, and education, because information is power. I was confident that the message that STEM careers were a path toward independence was not being amplified enough to younger women, early enough.

My graduate studies were all done remotely, other than a few days or weeks a year onsite, one for safety training during my master's work, a few hours away in North Carolina, the other for residency each summer during my doctoral studies in Virginia, which was at first one state, then MANY states (and time zones), away. In the middle of my second year as a graduate student, I shifted from the East Coast to

San Diego, California, to become a grand nanny to my new grandson while continuing my graduate studies. In 2010, number two joined the family, and I nannied full-time until I started working on my PhD in 2012. After one return to North Carolina to collect the belongings I wanted, we finally sold the North Carolina property, losing all our equity, and finalized our amicable divorce. We each kept our pension and benefits. California became my new home because that was where my family was.

Even before starting work on both my master's and doctorate, I had gotten more engaged with the Society of Women Engineers (SWE) in the first decade of the new century, eventually at the society level since there was no section (what many call chapters) locally in many places I had lived over the years. Over the 25 years I had been a member of SWE, my focus was always on advocacy — reaching out to girls and educators to talk about engineering (and other STEM fields) as a future possible career for them and a way out of poverty and dependence, and that carried over into my engagement with them, including a few years as chair of both the outreach and program funding committees.

In 2010, with the ready availability of a LOT of social capital in the form of members attending the annual conference, the concept of a one-day collaboration was implemented by the outreach committee to bring together those role models with local girls to do hands-on activities sponsored by media and industry funds. The initial session in Orlando initially drew a few hundred girls, peaking at over 600 in 2016, then settled back to several hundred each year on the Saturday after the professional conference ends. Several thousand girls have participated since the first in 2010. My graduate work focused on the surveys conducted at the start and end of events such as these, and what they are telling us about the interest of girls and what we can do to pique it more.

What became very clear was that the critical component was immersing girls in a community of supportive people who looked like them, doing things that most girls were discouraged from doing by family, friends, and social media, triggering the birth of the SWENext program. SWENext was a way for girls to explore what SWE had to offer before they got to college. It provides students with leadership opportunities and SWE members with an outlet for ongoing engagement with the local community beyond the occasional outreach event. Now in its 10th year with over 5,000 members, its growth speaks to a continued need for attention to engaging girls in STEM before high school so that by the time they get to college applications, they are clearer on their STEM possibilities.

Shortly after I earned my doctorate, someone introduced me to the American Association of University Women's Tech Trek program, a non-profit project founded in 1998 to immerse STEM-interested middle school girls for a week on a college campus with inspirational women in STEM fields. By 2018, over 800 girls attended camps across California each year, and the program had expanded to several other states, with over 14,000 participants since 1998.

Initially, I signed on as camp director for the local San Diego camp, but COVID snuck up on us as we were getting ready for the 2020 camp season, and we pivoted to delivering a virtual camp. Between March and July that year, in collaboration with Qualcomm's Think-a-Bit Lab folks, we designed and delivered a half-day robotics course to 40 girls using Arduino kits and ZOOM. Materials were ordered, volunteer staff and speakers engaged, students invited, and we were off as a pilot virtual camp program.

In 2021, it was expanded to a full day, with speakers each day after lunch, followed by different STEM workshops in the afternoon, including DNA extraction, and building circuits and equipment like spectroscopes and microscopes. Evenings are more casual, with workshops and speaker panels, such as being prepared for STEM studies and careers, learning how to speak up about things important to you, and finding your tribe.

For most of my volunteer communities, being online was new territory. In 2020, it was new to the kids, too. Fortunately, for me and many of our collaborators who had been working remotely since the internet first surfaced back in the mid to late 80s, the main thing that changed was accessibility due to bandwidth issues or hardware availability via Amazon, such as speakerphones and webcams. Between my professional work and my volunteer work with SWE, web meetings and email were just tools you used for mass communication, although before 2020, fluency in using them was less relevant.

While most of my SWE collaborators were very tech-literate, it was likely because the average member age was in the mid-30s, and they were mostly engineers, educators, and technologists who used it regularly; the average age of my AAUW collaborators was 65+, retired, and from fields that did not rely on tech as much. Fortunately, between us, we were able to not only deliver that year but deliver it bigger and better to nearly 650 girls across California in 2021 and another 300 in 2022. The virtual camp was added to the portfolio permanently in 2023 as an inclusive option for girls who cannot otherwise join in person, for whatever reason.

After the camp season ended in 2022, I took on the role of Tech Trek program director. One accomplishment I take a lot of pride in is saving more than a few trees. Before 2020, the camp paperwork was just that – paper. COVID kick-started the process, but it took until 2024 before we were fully transitioned from 100% paper to 100% online for both personnel and financial records. As a private, non-profit, volunteer-run organization that provides support to minors, the state of California is very attentive to privacy and records regarding those minors, AND the background history and training of any adults who interact with them.

At the same time, we struggle with staying in touch with those same minors over time as they transition from middle school to high school, then college and work, which makes it more challenging to continue STEM, encouraging them along the way. Now heading into a new program year with nine camps running well, the current program focus is to actualize a Tech Trek prototype of SWE's SWENext community. Being surrounded and supported by people cognizant of and empathetic to the issues that women in STEM still face is key to survival in a STEM-dependent future. Part of that process involves us developing a thicker skin than the one that we are generally NOT encouraged to develop, but in a safe space.

Several Tech Trek alum groups (TTAG) have sprouted organically over the last ten years in local communities of dedicated AAUW branch members and program graduates, but there are still few - less than one out of four branches have them. In 2023, the committee published a handbook to guide starting a local group and formally established a Tech Trek alum community (TTAC) advisory board to help connect the alums who would not have the benefit of a local TTAG. More recently, funding was set aside to encourage further development of localized alum communities and to create and deliver impactful alum pre-college programs and tools to help navigate their STEM future.

Then, just as we were getting ready for camp in 2023, I was also diagnosed with Stage IIa breast cancer and underwent an arduous but successful regimen involving six months of chemotherapy, followed by surgery and radiation, then another six+ months of chemo. The chemo side effects were less than desirable, with debilitating peripheral neuropathy in my lower legs and feet that has settled into what may be permanent tingly, uncomfortable stasis, with me not being able to tolerate wearing at least half of the shoes in my closet for more than 30 minutes, sometimes less. This reality may also be unfortunately coupled with being one of the few who never get the same head of hair back that they had before, rather just light fuzz, 90% thinner than it was before treatment. One year after my last chemo session, my nails have started to come back, so I have accepted that recovery from toxic poisoning is a slow process, especially given that I am now in my seventh decade, when molecular regeneration puts the brakes on more dramatically.

What kept me going through the treatment process was that I had so much more to do. It may have slowed my stride, but my lifetime goal of seeing a critical mass, a tipping point, of women in STEM fields never changed, and I kept and still keep going. Fortunately, most of my work was still remote, so I was down physically but not mentally. One of the benefits of the unsolicited experience was that it forced me to humbly acknowledge my inability to superwoman my way through it and to ask for help. 'Chunking' more of my volunteer work tasks such that I could more easily delegate them to others in the communities, while staying responsible to see that the tasks get done, worked for everyone involved.

As in pregnancy, I was forced to cede control of myself, only this time to trust the science that would take me to the brink of death, then bring me back to kill the runaway cell growth in me. Or at least that is what it felt like during the final treatment willingly undertaken, knowing it would get worse before it got better. The treatment worked, and at a cost that I would pay again. Would that the 15+ years of cellular aging that it does organically could be reversed as simply, no? Just another puzzle for an enterprising young woman to solve.

But to do that, she is going to need GRIT. She will need Guts with a capital G to demand a seat at the STEM table after sharpening her elbows so that she can make SURE she is heard, particularly in engineering or technology-based fields, which include most of the higher-paying occupations worldwide. Over time, she will build Resilience, the ability to bounce back from failure and rejection, which seems to waylay many women on STEM career paths. Many of us take it all too personally. It is not about us, and maybe that is what bothers us the most. The loss of the human factor.

Everyone is entitled to some downtime after a defeat, but the briefer, the better. Bouncing back from failure often needs a catalyst, something that might nudge us, forcing us to take *Initiative*, which is the third element needed to develop GRIT. Doing something that challenges our brain and puts us back in learning mode, providing the momentum to move ourselves to the next rung in life, whatever it may be, enables us to persist on the journey and renews our mission. Finally, perhaps more important than the other three, is that no matter how many times you fail, or how many shortcuts that become wrong turns you might take on the path of life, you must develop the *Tenacity* to never give up on succeeding in any endeavor, but especially if it involves group dynamics where you are not necessarily part of the "in" group.

After 50 years, I am discouraged that bigger strides toward parity have not been achieved for women and others who are less represented in STEM fields, particularly technology and engineering. Unlike the live frog in the pot of cold water being brought to a slow boil that does not realize what is happening until it is too late, I jumped out of the pot with enough momentum to evaluate what change has taken place in the last quarter century, and the answer is clearly 'not enough'. Although females may represent 30% or more in certain STEM fields, often, it involves occupations dealing with people rather than things.

Fewer than one in four engineers is female, when we are as good as males at problem-solving, math, and science. Why does that persist?

The next few generations have their work cut out for them to solve the problems that earlier ones created, but finding a safe space to share your inner self with all its geekiness is still a requirement for girls and women seeking longevity and self-actualization in a STEM field. Our biology does play a part in decisions we make along the way, voluntarily or otherwise, but there is strength in numbers, and collectively, we can make an impact. I think it was Malcolm Gladwell in the book with the same name who suggested that when the 'out' group gets to the 35-40% mark, that is the 'tipping point' of organizational or societal change. It would be nice to get there in my lifetime for women in STEM, but it may be in my granddaughters' instead.

Connect with Dr. Mary:

A Rising Tide

Coauthored by Amanda M. Tijerina
Senior Engineering Consultant

Where It All Began

I was born in Kenosha, Wisconsin, located midway between Chicago, Illinois, and Milwaukee, Wisconsin. A blue-collar town formerly centered around the automotive industry. Both of my parents were in the military when I was a kid, so my younger sister and I spent a lot of our childhood living with our great-grandparents. Our great-aunts and our great-uncles would pick us up to do things with them. We would do a variety of activities depending on the time of year. We learned how to fish, how to do yardwork, woodwork, paint, and volunteer (at the Spanish Center, St. Thomas, and St. Vincent De Paul). Everyone was handy with fixing and repairing things, and someone was usually working on a car. This is where my tinkering spirit was initially fostered. As a visual learner, this early hands-on introduction to being able to touch and manipulate something made it easiest for me to understand how those project items work and observe them in action. My Dad must have seen this, because he registered me for a multi-week STEM summer camp at a local college when I was in sixth grade. This was the first time I saw students and educators who looked like me. The diversity of students put me at ease because everyone was automatically accepted for who they were and how they looked. I remember every session being engaging and exciting. Also, all the students were having a blast learning, which was different from school. While I was walking through the halls between sessions, I knew I would go to college or a university. It felt right and possible simply by being in the atmosphere, which stuck with me.

The great-grandmother who helped raise me endured a lot, having immigrated to the US from Mexico as a teenager. Her fighting spirit and bold yet gentle presence left a lasting impact on me. She was creative in finding jobs to make a living while raising her family of seven kids, never afraid to get her hands dirty or do the work that others could not or would not take on. Her resolve pushed her to find a solution to all obstacles in front of her, even educating herself at night. My great-grandmother was one of my biggest cheerleaders, reminding me and showing me that anything I set my mind to is achievable.

I knew from a young age that I was going to leave my hometown to experience more of the world. I remember being in the backseat, looking outside at the landscape or moon with my forehead pressed against the window, imagining a time and place beyond my hometown and the current moment. Without fully understanding what I could achieve in my life, I set one of my first goals as leaving Kenosha, taking the determination I had learned from my great-grandmother, and the tinkering spirit I had learned early on from my extended family, and charting a course to get into a good university and unlock as many possibilities as I could for myself. I set my mind to earning and achieving straight A's and being active in diverse extracurriculars, all while working part-time jobs to pay my own way.

The Road To Engineering

When I was a junior in high school, I asked my favorite math and science teacher what I could major in to unlock as many opportunity paths as possible for myself. I had no role models or mentors who had gone to university, nor focused their career on math and science. I explained I knew I was good at math and science and enjoyed those subjects, but did not want to work in a hospital setting. He asked me if I had considered engineering, and I promptly responded that I did not know what engineering entailed. He gave me the assignment to go to the library, explore the various types of engineering, and report back to him what types of engineering were of interest to me and why, along with any outstanding questions. I was most intrigued by nanotechnology, industrial, and mechanical engineering.

Mechanical Engineering was ultimately my decision due to its hands-on, physical, and tactile subject matter and the flexibility of industry that the core principles offered. Additional factors I examined between my initial goal setting and finally getting my degree were the breadth of subject matter content, the job placement rate following undergrad, and the breadth of job opportunities across many industries.

After a year-and-a-half journey of determining which major and where to undertake my undergraduate degree, I moved into the 3rd floor of a dorm on Wisconsin Avenue in downtown Milwaukee, Wisconsin, on Marquette University's campus. I was placed on the women's engineering floor, where I developed some of my longest-standing friendships and relationships that I had not thought possible by simply living on the same floor as someone. Having endured some of the same headwinds and challenges with these outstanding female engineers over the years has allowed us to offer support, guidance, and a safe, built-in space to connect as we have continued to progress through our careers in this still-male-dominated industry. Additionally, with these engineers, I joined the Society of Women Engineers (SWE) in my freshman year. SWE has consistently offered me community, a sense of belonging, and purpose over the past twenty years in every place I have lived.

Marquette was the start of a new adventure and chapter that I knew would lay the foundation for years to come. As a Mechanical Engineer, I worked hard to make it. I applied for dozens of scholarships, grants, and financial aid options, and worked multiple jobs. I elected to do the five-year cooperative (co-op) student program, which is where I applied and interviewed with companies to work full-time every other semester for the duration of the program. Marquette has a strong industry connection and utilizes this to develop an excellent co-op program. The data showed job placement following undergraduate increased with job experience during undergrad, so I knew the co-op program would be the best fit with my goals. I would graduate with a few years of job experience, earn a decent income while co-oping, and learn what I enjoyed and disliked doing on the job as an engineer. I co-oped with General Electric Healthcare (GEHC) in the following positions: Systems Engineer, Technical Marketing Engineer, Sourcing Engineer, and Lean Manufacturing Engineer. While I originally saw myself working at GEHC long-term, starting as a co-op, onto a rotational leadership program, and then on through to retirement, I unfortunately did not get into the rotational leadership program I had set my sights on, and I was devastated. It was 2009, and the "Great Recession" had started, so there were hiring freezes and many companies, including GEHC, were laying off entire teams. The writing was on the wall, so I had to find a new home with another company.

Career Journey

I hold a Masters of Science and Engineering (MSE) in Engineering Management and Leadership from Purdue University and a Bachelor of Science (BS) in Mechanical Engineering from Marquette University. I have spent my career working across multiple highly regulated industries (medical devices, aerospace, defense, semiconductor, and financial services). I have built my career on being a highly versatile leader with experience leading cross-functional teams to deliver quality technical solutions in lean, fast-paced, and customer-oriented environments. As both an Engineer and Program Manager, I led teams and customers across multiple countries and cultural boundaries, while maintaining my passion for guiding and mentoring others to see their full potential.

As I was preparing to depart Marquette with my BS in Mechanical Engineering, I utilized my connections through the Marquette SWE chapter to get an interview with one of my dream companies, working on something that "pre-Marquette me" could have never imagined: I received a full-time offer to work at the Boeing Company as a Product Review Engineer.

While at Boeing as a Senior Product Review Engineer, the most memorable event I have is being on the tarmac witnessing the first flight of the aircraft I had worked on for several years, the Boeing 787 Dreamliner, the first fully composite commercial airplane. After leading multiple major design changes, it was extremely rewarding to see this aircraft lift from the runway, and the landing gears stow away. It was an exhilarating and emotional milestone to be a part of an enormous program working on high-impact projects.

After several years at Boeing, I made a change. My husband was relocated to California for his job. While there are many opportunities afforded to continue to work in Aerospace, I chose to open the aperture and look for a company that could offer me the most autonomy and the most impactful position. With that in mind, I took my next role at Intel Corporation as a Program Manager, leading the Lean transformation across the system validation and test labs globally. I was responsible for standardization, commonization, and training lab technicians and engineers on Lean methodologies and how they dovetail with

Agile methodologies. To do this successfully, I needed to build working relationships with individuals and teams across the 38 labs at 11 global sites. I began visiting and meeting with lab teams, gaining feedback on what is working, why something has not worked, and identifying areas of improvement. By listening to all the labs, I synthesized the data and showed them they had more in common than they realized, despite the labs spanning geographical, cultural, and religious boundaries. From there, I drove the creation of a common first silicon power-on process, which improved efficiency and throughput across the labs. This experience drove home a message that I had always known intuitively, but had not seen played out quite this way before: We all have something in common; we just need to be open enough to discover it.

After Intel, I pivoted to Discover Financial Services and then back to the aerospace industry with Gogo Inflight Entertainment, working for each as a Program Manager. While at Gogo as a Senior Program Engineering Manager for Aircraft Integration, I created a new product development and launch roadmap for key links within the satellite communication (SATCOM) systems. I built these from scratch and initiated verification and validation (V&V) activities within months.

After relocating back to California, and a short time working with Northrop Grumman, I decided I wanted to focus on the technical rather than programmatics; I returned to my roots with the Healthcare industry, working at Becton Dickinson (BD). While at Becton Dickinson, I was identified as an Inspire Quality (IQ) leader for global compliance to 21CFR820.60 Identification and 21CFR820.65 Traceability and recognized with an Inspire Quality award for architecting 24 procedures and driving global implementation on an IQ Blitz team.

Looking for greater flexibility in my career, I currently work with a boutique consulting firm, AME Quality and Compliance Solutions LLC (AME), as a Senior Engineering Consultant. I have always wanted to work for a consulting firm, and AME fits my overall life goals at this juncture in my life journey. This role draws on all my previous experience and enables me to have the opportunity to use the breadth and depth of my skillset, which keeps me challenged and engaged. This role has rewarded me with the ability to work with clients on a variety of products, services, and programs in different capacities. It is refreshing to flex a range of different skills while working from client to client, such as creating design requirements and drawings, writing procedures and technical documents, auditing systems, performing V&V activities (e.g., test), meeting with customers, and being hands-on with the product. I enjoy learning about various products and services, and it keeps me engaged and informed on what is happening within the industry, so I may best serve my clients.

Most recently, I was nominated for the Society of Women Engineers (SWE) San Diego (SD) section, 2025 Extraordinary Engineer Award by the fiscal year (FY)26 SWE SD President, Dr. Tracy Nguyen, and honored at the annual banquet.

While I am very proud of these individual career highlights, I am most proud of the career pivots I have made, where each transition I leverage what I have learned from previous experiences to make a large impact in a short period of time. Since my days at Marquette, I have worked to develop a strong core of problem-solving, breaking down a problem into components, and developing plans of attack for each component while maintaining a larger understanding of how everything works together. Demonstrating how my skills are transferable and bringing diversity of experience to each new opportunity is what continues to bring me to high visibility and critical projects for each company I have worked for.

A Rising Tide Lifts All Boats

From my early days at Marquette University, I saw the need for women to help each other in engineering and seek out others who understand that one woman's rise is not at the cost of the woman next to her. My cadre of college freshmen on the 3rd floor knew it then, and many of us continue to work to ensure that as we rise individually, we help to raise the tide for all other women in our industry. This guiding principle of "a rising tide lifts all boats" has foundations in social justice and economics, but I have found this idea to be a stark contrast to others who may view women in engineering as a zero-sum game where the advancement of one means that others will not be able to advance because that role has already been "taken". With this in mind as I progress in my career, I have been a constant presence in SWE and a persistent mentor to others around me informally and via SWE avenues, such as outreach and conference sessions.

My favorite way to give back is through outreach, such as a STEM camp professional mentor (e.g., Washington Aerospace Scholars (WAS) at The Museum of Flight, Seattle, WA), STEM workshop volunteer (e.g., table STEM experiments and activities), resume and career building session mentor (e.g., career night at a local college or university), and conference session speaker (e.g. SWE and Society of Hispanic Professional Engineers (SHPE)).

Finding joy in helping others navigate their path and advocating for those without a platform has led me to projects in unlikely circumstances. For instance, my passion for health and serving others, especially underrepresented groups in STEM, was strengthened during my parenting journey after struggling with infertility and finding a lack of information and support at a previous employer. Through extensive research, I have built comprehensive tools to merge and analyze all benefits (employer's, state, and federal), necessary actions, and planning to successfully go on Parental leave. I realized: if I was having difficulty finding this information, what about the other 70,000+ employees internationally in the company? How many others were experiencing this struggle across the STEM community globally? To foster equity and belonging for new working parents, I created parental leave templates that streamline the intricacies of going on and being on leave, which were presented to Human Resources (HR) Diversity, Equity, and Inclusion (DE&I) leadership. Areas of opportunity were identified for future implementation by the company's Total Rewards and the 3rd-party benefits coordinator. As HR continued their evaluation of adoption (including people manager training), the parental leave guidance and templates were shared via community channels to support individuals, including the Women's Initiative Network (WIN) Associate Resource Group (ARG). I still saw an opportunity to share my learnings with the broader STEM community. This motivated me to present at SWE's Local 2024 Conference in Las Vegas, "How to Plan & Communicate Parental Leave Transition." Then, upon my return to work, I was presented with having to manage breastfeeding in unsanitary and uncomfortable environments. I identified the lack of equality, equity, and sense of belonging presented in these lactation spaces. As the WIN+STEM Global Lead for Engagement and Belonging (E&B), WIN+SWE Global Champion Branding Lead, and WIN-SD Parenting Circle member, I knew I could and I needed to drive change immediately. I interviewed dozens of associates globally to understand their current state of local/regional lactation facilities and their respective experiences. After analyzing the data, I presented the findings to HR DE&I, identifying some sites (including my San Diego site) as non-compliant to local regulations, becoming a thought-partner with HR and Facilities around improving lactation spaces globally. I identified the program sponsor as HR DE&I, who accepted the sponsorship, ensuring each site's and region's HR Business Partner (HRBP) and Facilities evaluate their respective lactation spaces against their local regulations and perform an assessment on gaps for improvement. I worked directly with my business unit's HR VP, HRBP, and local and regional Facilities to build an additional lactation room at the San Diego site, where I served as the liaison between HR, Facilities, and users to address their needs in an efficient and timely manner. As a direct result, overall cleanliness improved, privacy measures were added, and refrigerators were installed to improve the health and safety of breast milk.

While this specific quest for gender and mother parity is unique to my experience, I share it here to recommend: Be the leader from wherever you are, attack the disparity that you see, and know that there are many others who will benefit from the changes you can drive. By fostering engagement and belonging among my communities, driving gender parity and work equity, while amplifying others' inspirational journeys, I have made invaluable impacts within my companies and communities, ensuring the voices of individuals across all intersectionalities are heard and taken seriously at all levels.

Reflection

While my career has well exceeded what I could have imagined as a young person, there are knowledge points I wish I had as I was climbing through the career obstacle course. One of the most painful lessons I have learned is that as I progressed in my career, it became increasingly challenging to find willing and genuine mentors and sponsors in influential positions and spaces that were interested in my best interest and truly valued my career aspirations. These relationships are invaluable in helping navigate career path options and viability, especially in constantly shifting markets and technology. Mentors and sponsors, especially in large companies, are often able to identify paths or experiences that may not be visible to you and me. To overcome this challenge, I found it beneficial and rewarding to seek and foster mentorship and sponsorship outside of my immediate work organization, via employee resource groups, professional societies, and peer mentors inside and outside of my current industry. This enabled me to see and access opportunities in different industries and different physical locations as I made my career pivots.

One other strategic piece of information I wish I had known about is the importance of self-advocacy from the start of my career, because just doing a great job is not enough to be recognized for a pay raise, promotion, or that position you have been working toward. In short, it is not who you know, nor what you do, but rather who knows you for your work. Had I known this sooner, I would have started advocating for myself earlier and differently. It may have prevented me from being frustrated and building resentment in certain situations early in my career.

Likewise, I wish I had known negotiation is essential in many engineering situations. Had I known, I would have taken a negotiation course and training earlier.

Before choosing engineering, I wish I had known that writing is the skill I would use the most. While I am now happy that Marquette forced me to take a writing class (not all universities do), I was under the impression that once I completed my English courses in undergrad, I would be done with learning how to write, compose written documents, and illustrate arguments creatively. All this being said, it may have swayed me to avoid engineering if I had known beforehand.

Lastly, I never could have imagined being where I am today – where I live and have lived in the world, the companies I have worked for and with, the career I have shaped, and the life I have built. I live in San Diego, California, with my husband and our two young sons, where we enjoy traveling, exploring, and being outside. If I could advise my younger self, and emerging STEM leaders (that's YOU), I would kindly recommend: Be open to new experiences. Say yes to opportunities that are advancing your skills, particularly beyond your comfort zone. Say no to opportunities that are not advancing your skills and holding you back in traditionally female, busy work, instead of strategic value-add work. If you see disparity, attack it. Listen to your gut and evaluate how an opportunity aligns with your values and goals versus a rigid black and white career plan that may not be all-encompassing. Every time I update my career plan, I have made it increasingly broader to facilitate exploratory conversations and keep the lines of communication open for potential opportunities from unlikely sources. I have found there is significant value in diversity of experience (think: each job role change is at least one sigma different from the previous), and it has served me well because with each transition, I fold forward what I have learned from previous experiences to make large, impactful contributions in a short period of time. If you take anything away from this chapter, remember you can do anything you set your mind to, because *where there's a will, there's a way.*

Connect with Amanda:

My STEM Journey: Why Mindset Matters

Coauthored by Aileen Fahme
Mechanical Engineer

I WAS BORN IN the Philippines, the second oldest of four, two brothers, one sister, and me right in the middle of the chaos. When I was eight, my mom made the hardest decision of her life. She moved to America to find better opportunities so she could give us a future she never had. For two years, it was just my dad, my siblings, and me taking care of each other the best we could. My dad ran a small tailoring shop, and that's what kept food on the table. Whenever my mom could send money, it felt like a little reminder that she was still fighting for us from across the world.

When I turned ten, everything changed. My mom was finally able to get our green cards approved, and she brought all of us to America. That moment completely shifted my life: new country, new chances, and the start of a journey I'll always be grateful for.

We lived in the Bay Area, in a town called Mountain View. In our small two-bedroom duplex, all four of us shared a single room as we grew up. We didn't have a lot, but we had each other. It was loud, chaotic, loving, and fun! Everything, a big Filipino family usually is.

My dad was the heart of our home. He was the type of person who would tell a joke he could never finish because he'd be laughing too hard halfway through. I loved that about him. I was especially close to him because he was always there in the small moments, like waking up before sunrise to help me deliver newspapers. It was quality time. We'd toss papers, complain about the cold, joke around, and then treat ourselves to two warm hash browns from Jack in the Box. Honestly, those hash browns were very delicious. But more importantly, those mornings taught me something I didn't recognize until years later: mindset matters, even when you're ten.

It wasn't about newspapers. It was about waking up early, working hard, supporting each other, and somehow still laughing through it. Without realizing it, I was learning the foundation that would carry me through every challenge in STEM (Science, Technology, Engineering, and Math).

Learning English and Finding Strength Where I Least Expected It

Starting school in the U.S. was… rough. My siblings and I could read and write English well enough, thanks to school in the Philippines, but speaking it? That was a whole different adventure. I had the vocabulary, but it lived in my head, not in my mouth.

So, we spent our first couple of years in ESL classes (English as a Second Language). It felt like everyone else was sprinting while we were still tying our shoes. But once the language barrier slowly faded, school started to feel less intimidating.

Because English felt overwhelming at first, I leaned toward subjects where talking wasn't a big part of the grade—math and science. These subjects didn't care about my accent. They didn't care if I used the wrong verb tense. Numbers just wanted you to focus. Physics just wanted you to think. They were predictable, logical, and stable. And the crazy part? I got good at them. Not because they were easy, but because solving a problem after struggling with it gave me this weird rush…like, "Wait, I actually did that."

And looking back, that was mindset at work. I didn't call it that at the time. I just knew I liked the feeling of figuring things out. But that's what mindset is: you start seeing challenges not as "I can't," but as "Give me a minute… I'll get there." For anyone in STEM, especially young women, this mindset is gold.

Music, Engineering, and A Very Big Bass

Like a lot of STEM kids, I had a huge love for music. In sixth grade, I ended up with the upright bass. If you've never seen a petite Filipina girl carrying an upright bass, imagine a hamster trying to drag a Costco-sized suitcase.

But I stuck with it. I improved. I even got nominated to play with a college orchestra. It felt huge, until my dad, in the most loving way possible, said something like, "Anak… are you sure you want to keep carrying that giant thing around?"

So, I switched to the cello…And immediately, I fell in love with it.

The cello became my emotional outlet, my stress-relief, my way of breathing when life felt overwhelming. When school was tough, music grounded me. When I felt doubt creeping in, the cello reminded me I could still make something beautiful.

And music, just like engineering, wasn't easy. Some rehearsals were brutal. Some pieces felt impossible. Some practices ended in tears. But I learned that loving something doesn't mean it won't challenge you. That's a mindset thing again. Passion and frustration can exist together. And when they do, that's usually where growth happens.

In high school, I joined the California Youth Symphony. We traveled to Germany and Austria, competing against some of the top youth orchestras in the world. It was incredible. It was intimidating. It was exhausting. It was life-changing.

Our music director was strict, and I constantly worried about messing up. But I kept going. Every stressful rehearsal, every sore muscle from playing too long, every moment of doubt, it all pushed me deeper into the mindset that later carried me through engineering: You don't have to be fearless. You just have to keep showing up.

Choosing Mechanical Engineering—Even When I Had No Idea What I Wanted

When I started college at UC San Diego, I chose Mechanical Engineering and minored in Music. People sometimes assume I had this perfectly mapped-out plan. Nope. Not even close.

The truth?

I picked mechanical engineering because my advisor told me, "Choose this for now, you can switch later once you figure out what you want."

Mechanical engineering became my home by default… and eventually by love. But not quickly.

College was a struggle. There were nights I cried in front of problem sets. There were classes that made me question everything about myself. There were days when I felt like the only one who didn't understand what was going on. And honestly, back then UCSD's SWE (Society of Women Engineers) was so quiet that I didn't even know it existed. I didn't have a big STEM community to lean on. I relied on a few friends, my family, and pure stubbornness.

And every time I thought about quitting, I thought about my parents, about all they had given up moving us across the world. I thought about those early mornings delivering newspapers. I thought about my younger siblings watching me; I thought about my mom and dad wrapping lumpias to sell for added income.

And somewhere inside me, a voice kept saying, "Keep going. You can do this."

That was mindset... not confidence, not magic, not genius—just mindset.

And it made all the difference.

Turning Theory into Reality: My First Engineering Job

After graduation, I joined the Engineering Rotation Program at Solar Turbines—a gas turbine manufacturing company that's part of the Caterpillar family. Through the program, I rotated across several departments: materials, mechanical design, development test, and combustion engineering. It felt like the "aha!" I'd been waiting for.

I saw how engineering lived in the real world, how real teams worked. How real products were designed. How real problems were solved.

After the rotation program, I decided to pursue my MBA at the University of Redlands. That opened the door to one of my favorite jobs ever: Project Application Engineer (PAE).

As a PAE, I worked with sales teams and customers to come up with solutions that would best fit their energy needs. I learned what customers actually cared about. I learned how gas turbines fit into global industries. I met experts from all corners of the company. I traveled. I ate food I couldn't pronounce. I saw how engineering affects people's lives far beyond spreadsheets and drawings.

I later became a group leader and had the chance to mentor 15 new Project Applications Engineers. Seven of them were women from totally different backgrounds. Getting to know each of them, what motivated them, what scared them, what made them feel alive, was one of the highlights of my career.

And seeing them go on to become Group Managers, Project

Managers, Sales Engineers, it felt like watching seeds I helped water grow into entire trees.

But again, it all came down to mindset. They had talent, yes. But the thing that pushed them into leadership was their mindset: the willingness to learn, to take feedback, to stretch themselves, to believe they belonged.

The Hard Moments (Because Every Journey Has Them)

Let me be real: not everything was sunshine.

Losing a project after months of late nights? That hurt.

Working with customers who didn't believe a woman could be an engineer? That hurt in a different way.

But here's where mindset mattered again. I didn't take it personally. I learned to keep my cool and stay focused on my mission: "Help the customer solve their energy challenges."

Not "prove myself."

Not "win them over."

Just do the work.

And once again, mindset helped me separate someone else's bias from my own worth. That skill is priceless in STEM, especially for women.

20+ Years Later—A New Chapter and a New Purpose

Over two decades, I've been a PAE, Project Manager, Product Support Engineer, Team Lead… you name it. I've worked with huge customers, led tough projects, mentored new engineers, and built lifelong relationships. Then, a couple of years ago, I decided to shake things up. I moved into operations, totally different from everything I'd done before, and became the Manager of Logistics and Shipping, leading a team of 12.

Let me tell you. Managing people is its own engineering discipline. People don't come with manuals. They come with personalities, hopes, fears, motivations, and stories. And I learned quickly that when you genuinely listen to people, support them, and help them grow, everything changes. They perform better. They feel better. They bring their best selves to work.

If there's one truth my career has revealed, it's that mindset matters more than any title you'll ever hold. Leading people showed me that real leadership isn't about having all the answers. It's about listening, caring, and helping others rise.

Why I Give Back and Why SWE Became Part of My Purpose

At some point, I realized everything I'd learned wasn't meant to stay with me. Someone once invested in me. Now it was my turn to invest in others. So, I reached out to SWE San Diego. The first person I met was Kelly Hon, the SWE SD President at the time. She told me they were looking for a Recording Secretary. I raised my hand immediately, not because I had the time, but because I had the purpose. I wanted to lift the next generation of women in STEM, especially young women who looked like me, came from immigrant families like mine, or didn't have access to role models growing up.

Over the next four years, I volunteered at outreach events, spoke on panels, mentored students, and supported as many young engineers as I could. Every single event reminded me why this work matters. When a little girl lights up because she met a real engineer? That's mindset changing in real time.

In Fiscal Year (FY) 25, I was honored to be elected President of SWE San Diego. Our team was unstoppable! We participated in more than 40 outreach events in the first half of the year alone, hosted fun member socials, and ran workshops that helped people grow.

At the FY25 SWE National Conference, I was awarded the SWE APEX Award for Advocating Women in Engineering. That award wasn't just mine. It belonged to every mentor, mentee, volunteer, colleague, and student who made that journey meaningful.

To Every Aspiring Engineer Reading This

My journey, from a kid struggling in ESL to a mechanical engineer, mentor, SWE leader, and advocate, wasn't built on being gifted or having all the answers.

It was built on a mindset.

On choosing to keep going.

On trusting myself even when I felt unsure.

On getting back up after setbacks.

On surrounding myself with people who lifted me higher.

Your skills matter.

Your determination matters.

Your voice matters.

But your mindset?

That's your superpower.

Own it.

Grow it.

Lean on it.

And one day, give it back to the next girl who needs to see what's possible.

Connect with Aileen:

Engineer Your Mindset

Coauthored by Natalia Pascual

To the next generation of engineers:

If [your engineering journey] is X, then X ≠ Y. In other words, your engineering journey is not linear. You may have learned this already. The late nights. The barely passing (or failing) exams. The moments when you wondered if engineering was even for you. Maybe some of you have even pivoted degrees or careers.

"Why did you do this to yourself, Natalia?" was just one of my many thoughts in college. But almost a decade later, through all the peaks and valleys, I can confidently tell you: it has all been worth it. I want to share a few of the lessons that I have learned along the way.

Let's begin at a time when I didn't even know the field of engineering existed.

Part One: You don't know what you don't know yet, and that's a good thing.

When I was younger, I wanted to grow up to be many things: I wanted to be a teacher, a detective, a dancer, just to name a few. I couldn't wrap my head around being just one thing when I had so many interests.

Your world is only as big as what you've been exposed to.

My world consisted of parents who taught theology, led Sunday school, and mentored people of all ages. Naturally, I wanted to be a teacher because I loved helping others. I loved the idea of leading a classroom and writing on a whiteboard. That was cool for little Natalia.

As I got older and started watching movies like *Spy Kids*, the gadgets and technology were fascinating! My world was expanding, and I now wanted to be a detective. My parents told me I could do anything I wanted in life, and I believed them. So being a detective was my new dream.

Then math entered the picture. At first, I was terrible at it; I struggled to pass my classes. Eventually, it clicked. The formulas, structure, and satisfaction of solving a problem made me love the challenge. I got confident enough in math that when my high school teacher announced a math tutor opportunity, I was intrigued. At fifteen, I had never thought about making my own money. I liked math, I liked helping people, and the idea of relieving any financial concern from my parents felt meaningful. Teaching math became my main source of income for the next eight years. I thought to myself, "Maybe I really will become a teacher."

It was about this time in my life that I discovered a new love. One night at a high school talent show, I saw Latin dancing for the first time. I was immediately captivated. I knew I wanted to be dancing onstage. When I tried learning to dance, I found out I had two left feet and was stiff as a tree. I took lessons, I practiced hard, and I joined the team and soon found myself performing and competing. For a moment, I wondered if becoming a professional dancer was a possible career choice for me.

While I loved dancing, I was never trained professionally. Sweet, innocent, and young Natalia had to accept that pursuing a degree in dance wasn't a realistic path for her. If I hadn't been exposed to teaching, or watching spy movies with gadgets in them, if I hadn't been exposed to dancing, I wouldn't have known those possibilities even existed.

As an engineer, one of your biggest assets is recognizing that you don't know what you don't know.

At this point in my life, I still had no idea what engineering was. No one around me was an engineer. My parents had immigrated from Guatemala, and I was a first-generation college student. While my parents planted the seed to reach for the stars, I didn't know which stars to reach for. Ultimately, I had to choose a major, for the sake of choosing. I picked applied mathematics. Maybe I could do something with this familiar field. But as I sat at my computer and I googled, "What can I do with an applied mathematics major?" I was introduced to another role: engineering. I learned that engineers create something out of nothing, engineers solve problems, engineers innovate, and engineers can help the world move forward.

I also learned that not many women, let alone Latinas, were engineers. Growing up, my parents nurtured another part of me: that driven, fierce side of me. I like to think it's a Latina thing. That side of me looked at those statistics and thought, "Unacceptable. I bet I can do that." Less than a semester later, I switched from applied mathematics to mechanical engineering. Even then, I didn't fully understand what an engineer did. It wasn't until I rewatched the movie *Iron Man,* and it was in the scene where he creates his first suit in the cave, that I finally understood what engineering was. He was so resourceful, innovative, and driven. It was during this time that I was taking a manufacturing class and realized that I was doing the same kinds of things (minus the flying suit). That's when I knew I was studying what I was meant to study.

There were many moments when I questioned myself. *Did I belong? Was I smart enough?* I felt like I didn't know what I was doing. But I had to remind myself that I don't know what I don't know. So, I kept embracing every step of my journey with an open mind. Today, choosing engineering has been one of the best decisions of my life. Was it hard? Absolutely, it was one of the hardest things I've ever done. Has it paid off? Absolutely, beyond what I could have imagined.

Part Two: Life is dynamic

Two things I quickly learned when I first started working in corporate America: The first one was that if I wasn't being challenged in my position, I became unhappy. Second, I was unhappy with the lifestyle of work, home, eating, sleeping, and repeating. In our Latin culture, especially as a first-generation graduate, you are taught to be grateful for what you have. Don't question it. Just accept it. I strongly believe in gratitude, but if I "just accept it," then how will I be challenged? I would think: *Could I have more? Could it look different for me?*

As an engineer, you are trained to ask questions, seek clarity, and challenge what is "normal" to innovate beyond bounds.

My first "big girl" job was as an analyst for a medical device company. It wasn't even in engineering.

Each day, I was doing manual and repetitive computer work. Due to the circumstances under which I was hired, there wasn't much innovative work to be done. I would ask myself, "Where was the challenge?" "Where was the growth?" I knew I had to start seeking my own ways of professional growth. The first skill I needed to strengthen was Microsoft Excel. I knew I had to improve many skills to make my job easier. I researched and learned how to create Smart Forms in Microsoft Excel. I was so excited with what I had created, and I knew I needed to share this with my manager. She was so impressed, and she had me share it with some of the project leaders, who loved the concept so much that they had me improve it for our global team.

While having key players on your side of the court is important (like my manager), it is just as important that you are your biggest advocate. If I hadn't shared the tool with my manager and just kept it to myself, she never would have known about it and shared it with the project leaders. If I hadn't shared that I was unhappy, she wouldn't have known how to help me. Part of being an advocate is voicing what you have accomplished and what you need.

From that point on I began working in tech focused projects. I had the opportunity to work on a project that involved creating a tool using machine learning (before generative AI). The tool won an Innovation Award! Being exposed to this new field helped me realize how much I loved it. I loved the beginning-to-end influence I could have on these projects! I loved the tangible outcomes that helped my peers, and I loved the innovative way of doing things. I loved it so much that I went on to earn a certification in Python and a master's degree in information systems management. Eventually, I became a Business Application Analyst and got to work on tech projects that help solve problems and make my team's work easier.

There was a part of me that felt sad about leaving my mechanical engineering dream aside because I thought I would go on to design medical devices. This was very different from what I had envisioned for myself. But I didn't know what I didn't know. I had to embrace the fact that life is dynamic. In engineering, dynamics teaches us about motion and the forces that cause it. In life, those forces take on the form of new experiences, new people, and new things we are exposed to, and that guide us in different directions. As you learn more about what's out there and more about yourself, that vision for your life might change. It might change multiple times throughout your life. I've learned to embrace these dynamics and the changes they bring.

Part Three: Engineering goes beyond your career; it's a mindset. Use it.

Engineering is not just a career you choose; it is a mindset you can carry into everything you do. We are trained to understand and research, ideate and design, and trystorm and iterate. That's simplified, but for example, at work, if we are given a problem to solve, we will:

1. **Understand & Research**: ask questions, gather insight, and get clarity on what the real problem is before acting.
2. **Ideate & Design**: envision possibilities, use critical & creative thinking, and shape potential solutions.
3. **Trystorm & Iterate:** try out those ideas, learn quickly from what worked and what didn't, apply the solution, and improve it over time.

Mindset is something you naturally carry through your life. I didn't recognize it at first; however, I had already been applying this mindset long before I worked in corporate America or became an engineer, and I'm sure you have too.

Engineering helped strengthen my way of thinking. I applied it as I explored new career paths, designed technical solutions, discovered my love for the field of engineering, and ultimately accepted a Business Application Analyst role to try something new. I didn't just use this mindset to advance in my career; I used it to expand my lifestyle. For example, in 2021, I was given the opportunity to travel abroad for work. My mind was blown that this was even an option for me! At this point in my life, exploring new places was already a big part of it, because I had started traveling in college. But I had never been paid to travel! So I was beyond excited!

Since I was already working remotely, I began to research the possibilities of remote working from Ireland. I designed a whole plan for what it would look like and advocated for myself by sharing the idea with my manager. Management agreed to let me try it out. I lived in Ireland for a month, and when I wasn't working, I solo-traveled around the country.

It was a lifestyle I didn't think I could have. While it wasn't a huge, complex technical problem that I needed to solve, I became fully aware that I engineered my life, and that was empowering. If I could figure this out, what else could I try? New experiences sometimes act as the force that helps you find motion again. I became more strategic in asking to be on projects that have helped me travel to Europe and across the US. On a personal note, I also strategically designed my travel plans and reached my goal of visiting all seven continents by age 30!

One year after I returned from Ireland, I founded Casa N, a space for Latinas in STEM to embrace connection through self-care, personal growth, experiences, and community. Casa N started with self-care retreats (aka getaways) to help burnt-out Latinas take a break and indulge in self-care. Today, Casa N is also on a mission to take Latinas to extraordinary places and experiences across all seven continents. As I've mentioned many times throughout this chapter, exposing yourself to new things expands your world. It allows you to learn more about yourself and deepen your connection with the world. In 2025, Casa N went on a retreat to the Arctic in Norway! (Maybe you will join us in the next retreat!)

The beautiful thing about all this is that there are no immediate results. Engineering quick wins or solutions is fine; however, the ones that take time to build, iterate, improve, and fully execute are the wins that matter most. As I look back on my journey, I see how I am blessed to have had a handful of those experiences: becoming an engineer, becoming part of the 4% of Latinas with a master's degree, visiting all seven continents, and starting a business to support Latinas in STEM (science, technology, engineering, and math). I could never have imagined being privileged enough to provide financial support to my loved ones or to take my parents on trips around the world. This past year, we even went back to Guatemala for the first time together!

In my eyes, it was all because I took a leap of faith and chose engineering.

I hope you notice that it wasn't the degree or career itself, but the skills engineering has equipped me with that have enabled me to tackle each point of my life with curiosity and an open mind. This doesn't mean it comes without struggles, mistakes, or wrong turns. But I believe engineering gives us the courage to trust ourselves enough to figure it out. Whether you have the engineering title for a short time or for a long time, the way you think is what makes you an engineer. It is the way you grow and improve; it is the way you question things and ask how they can be better. You have the right mindset to figure it out. That doesn't mean you have to do it alone, don't forget that! You are the next generation of engineers and will build amazing new things that will help our world. But it is important to remember the world you came from. Build for your family or loved ones, build for yourself, and build for your community. As the next generation of engineers, use this mindset beyond your career. Use it to elevate yourself, your life, your environment, and those around you. You've got this!

I believe in you.

Con Cariño,
Natalia Pascual

Connect with Natalia:

Instagram: @soynataliapascual

SOYNATALIAPASCUAL

This Is How I Roll: A Metallurgical Engineer's Journey Through Manufacturing and Community Advocacy

Coauthored by Vanessa Li Davis
Lab Leader

Manufacturing Roots: A Seed Within Me That Took Time to Grow

My name is Yin Wai Vanessa Li-Davis. As a kindergartener, I was fascinated by science and space. I imagined myself working with rockets. I loved watching chemical reactions and physics experiments as they were hands-on. I didn't like sitting still, and science let me move, explore, and discover. That curiosity never left me.

Growing up in British Hong Kong, my younger sister and I attended an all-girls school established by the Anglican church, a school steeped in tradition and focused on music and home economics. I took piano and flute lessons, learned needlework (which still comes in handy every now and then!), and culinary arts. My parents do not have a science or engineering background, but they have always encouraged me whenever I tell them about how fun chemistry and physics are. Back in the 1990s, the terminology STEM did not exist; we simply referred to this as science.

My grandfather was a managing partner of the Oceanic Cotton Mill in Hong Kong in the 1960s. He later opened a household appliances factory in Aberdeen, Hong Kong. Although my father has a business degree and no background in science or engineering, he often researched and developed new products with my grandpa, such as popcorn machines, toasters, mixers, hair dryers, and curling irons. He would try to improve product designs in order to raise productivity, reduce material costs, and enhance their cosmetic appeal. I found these appliances very interesting, especially since I had not seen a popcorn machine other than in the cinemas. Popcorn machines are not a popular household item in Hong Kong.

I remember visiting grandpa's factory in the 1990s. My curious mind thought everything was fascinating, and it was the best playground in the world!

My dad was the membership director of the Federation of Hong Kong Industries, an organization similar to the Chamber of Commerce. Sometimes he would tell my sister and me about "Mr. X decided to move his toy factory across the border to mainland China." Toy manufacturing is a very challenging industry. It is labor-intensive and involves manufacturing expertise in plastics injection, metal stamping, spray painting, sewing, and stitching (making garments for dolls). The many stories that my father told us were captivating. These stories had indirectly affected me. It started planting seeds of what I would do in the future, although I did not have a clear career goal in mind at the time, nor did I know what engineering was about.

Teachers, Transitions, and Taoism: How I Found My Path

In Form 1 (equivalent to 7th grade), my physics teacher, Ms. Yau, challenged us to design either a thermal cup or a tissue paper that wouldn't tear after being soaked in water. My group proposed a stainless-steel cup with a vacuum layer—this was my first exposure to material selection and design. At the time, I hadn't even heard of materials engineering.

In Form 3 (equivalent to 9th grade), Ms. Yau brought a small, beautiful optical fiber Christmas tree to class and asked if we knew how it worked. Dial-up internet existed but was expensive, slow, and not widely accessible in the 1990s. I spent time at the library researching what made the tree so mesmerizing. That was the first time I learned about optical fiber. I still had not heard of materials engineering.

Interestingly, when I reconnected with Ms. Yau in 2023, she told me she hadn't heard of materials science as a major either, so I ended up introducing something new to the teacher who first sparked my curiosity about materials. That moment felt full circle.

In Hong Kong, we had to choose between art, science, and commerce during Form 3. Science was no doubt the choice for me. There was a teacher who constantly reminded us that science is not for everyone. "If you are not good at math, then science may not be suitable for you." This statement discouraged girls from choosing engineering and undermined those with potential. Later in my life, I discovered that this statement is a myth! In addition, there are a wide variety of engineering disciplines. Not all of them require advanced math knowledge. Most importantly, not doing well in one math class is not a sign that you will fail as an engineer in the future. Nowadays, there are lots of tools and software that can help double-check the accuracy of your calculations.

At 16, I left Hong Kong to attend a boarding school in California. It was a big change. As a boarding student with limited internet and phone access, I got homesick during my first semester. I spent most of my time reading—mostly physics books. Originally, I planned to study mechanical engineering. My father had a chance to meet with Nobel Prize recipient and Father of Fiber Optics, Sir Charles Kao through an industrial business trip in Finland; he introduced him to materials engineering. I had never heard of such a study before, but it involves chemistry and physics—my two favorite subjects—so I gave it a shot.

I enrolled in the Materials Science and Engineering (MatSE) program at the University of Illinois. It was a tough major—only about 50 of the 70 students in my freshman class made it to sophomore year. Despite these statistics, I never considered switching as this was my interest. During my freshman year, I spent a lot of time studying. Some of my classmates prioritized other activities instead of studying, and at first, I thought they were smarter than me, or they learned more quickly. Fast forward to my junior year, and I noticed two of my classmates had "disappeared". I later learned that they were dismissed from the school due to having poor grades. I then realized it wasn't that I was not smart enough, and my hard work was worth it. I also realized that it is okay not to be the smartest in the class or in the meeting room, but one must be well prepared, as opportunities only come to those who are ready. Growing up, my mom always says "沒有不勞而獲. Never expect reward without putting in the effort." I did not have a clear picture of my career back then, but I was clear that I needed to put in hard work.

The MatSE program at the University of Illinois required us to choose a concentration among biomaterial, metal, polymer, ceramic, and electronic during our sophomore year. To decide, I used a mix of process of elimination and Taoism:

- Process of elimination: Biology was okay and not my favorite. I did not plan to go to medical school. Electronics was okay, but that professor wasn't my favorite. Then, I was left with ceramics, metals, and polymer to choose from.

- Five Elements (五行): Ancient Chinese alchemists and Taoist philosophy believe that there are deep connections between humans, the universe, the ecosystem, and the landscape—patterns that could even be understood through observation and statistics. In this framework, polymers are associated with fire, metals with metal, and ceramics with earth. A Feng Shui Master once told me, based on my birth date and time, that I needed more "water". Since "metal generates water," I chose metallurgy as my field of study. Believe it or not, that's how my career path began.

Now, the real elements: My Journey in Iron and Aluminum

My first adult exposure to steelmaking was a plant tour at a steel company during my sophomore year of college. I did an internship at another steel company and did statistical process control work with their Electric Arc Furnace and tundish data. Early in my senior year, my professor discouraged me from pursuing manufacturing as he said, "Steel mills are dirty". Well, it's true that some areas such as the blast furnace could get dirty; however, automation and new technologies are improving the situation. I found out throughout my career that the environment can differ based on manufacturing units and how well housekeeping is kept. As I look back, I am glad that I persevered.

Internship - Polymer Company X

After my sophomore year, I completed an internship at Polymer Company X in Hong Kong, where they specialize in developing polymer resins. I worked on tensile testing of different types of plastic at a local vocational school laboratory, then shared results with product engineers. During the internship, I also attended a trade show in China, where I observed how product engineers effectively combine technical expertise with strong sales skills when introducing products we offer to potential customers. This is where I learned that technical knowledge is foundational and your degree opens doors. But leadership is defined by your ability to influence people and solve problems, such as boosting productivity, reducing waste, and streamlining the workflow. We also need to be business-savvy and be able to tailor our message to our audience, which is important. Before selling an idea or products, listen to understand your customer's needs and put yourself in others' shoes. These skills will elevate your impact far beyond technical contributions.

Internship - Steel Company Y

After my junior year, I completed an internship at Steel Company Y, where I completed data analysis related to hot straightening and its impact on production yield and cost of goods sold. I also assisted in the Split Heat Grind Trial, which involved product testing for both continuous casting and electric arc furnace processes. This experience deepened my understanding of Statistical Process Control (SPC) and process improvement. It also gave me valuable technical experience to discuss in future job interviews, beyond just behavioral questions.

Be strategic at the job fair

At the start of my senior year, I attended the campus job fair where companies came to recruit students. I reviewed the list of companies attending in advance and focused on the top three that aligned with my interests. One of my target companies was Steel Company 1.

I spoke with an HR representative from Steel Company 1, who scheduled an interview for the following day. What stood out about this interview was the balanced mix of technical and behavioral questions—roughly 50/50. At the time, I was a bit shy and thus not confident with behavioral questions. However, I felt much more comfortable when the question was about my internship experience. Having concrete and data-backed examples helped me speak with confidence and clarity. I was later invited to attend an on-site interview. (In the 2000s, video interviews were not popular.) I took a Greyhound bus up to Northwest Indiana for an interview, as I did not know how to drive. My advice for international students is to obtain a driver's license before you graduate.

I highly recommend enrolling in a driving school, as a professional instructor can speed up your learning process.

The day began with a safety video and Personal Protective Equipment (PPE), followed by a tour of the blast furnace and an interview with the operations team. The afternoon was spent in the tin mill and an interview with the quality department. It was a full eight-hour day of interviews and a walking tour. During that visit, the steel-toe shoes provided for visitors were men's size, and by the end of the day, my feet were covered in blisters. I later learned to ask about PPE requirements before on-site interviews. Despite that, the experience proved incredibly valuable, giving me a strong understanding of the company's operations. In October—seven months before graduation—I received and accepted the job offer.

[Advice for those looking to pursue a master's degree: You can start your career with a full-time job and pursue your degree part-time. Many employers offer tuition reimbursement as long as you maintain good grades, letting you earn while you learn. If you leave, you may need to repay a pro-rated portion back, but a future employer might offer you a sign-on bonus to offset that cost.]

Early career - Steel Company 1, Process Quality Engineer HDGL, Northwest Indiana

Don't wait until graduation is near to start applying for full-time positions. Securing a role early gave me peace of mind for the rest of my senior year; I could focus on passing my remaining classes knowing a job was waiting for me.

My first full-time position in manufacturing began at a major steel producer in Northwest Indiana (Steel Company 1), where I worked as a Process Quality Engineer supporting the Hot-Dipped Galvanizing line (HDGL). During my first week of work, I was navigating my way as a newly licensed driver in a different town while I was navigating my new job and employee benefit options. I wish I would have taken time to learn about 401(k)s, IRAs, and HSAs prior to graduation. That way, I would not have to rush during my first month of a new job. As an immigrant, I was unaware of the tax advantages of a Roth IRA. I missed the opportunity to contribute to my early working years.

Steel Company 1 is over 100 years old and was very well established. This was a great learning ground as lots of colleagues have over 30 years of work experience. They taught me a lot.

This experience became the foundation for my technical discipline and problem-solving mindset. I learned about welding, the cleaning section (concentration and conductivity), continuous annealing furnace (different steel grades, line speed, soaking, peak metal temperature), zinc pot and air knife (coating weight), post treatment (chemical coating), skin pass (surface finish and elongation), tension leveling (flatness), electrostatic oiler (lube weight), inspection, and testing. Collaborating with operations, field service, and scheduling teams, I helped troubleshoot issues, implement corrective actions, and drive continuous improvement. I saw the power of data-driven decision-making and the importance of aligning technical solutions with real-world production challenges, especially in how our processes directly impacted customer applications (painted building panels, appliance panels, and automotive).

Aside from the technical and industry knowledge I have gained from this experience, I also learned the importance of embracing feedback. I recommend you align early whenever you start working with a new manager or new team. Don't wait for the annual performance review. Proactively seek feedback to uncover blind spots and growth opportunities. There is a Chinese proverb: "金石良言" (Wise advice is as precious as gold and as enduring as stone)—meaning that feedback is a valuable gift. When someone offers constructive input, they're stepping out of their comfort zone to help you grow and succeed.

Layoff Isn't the End—Something Greater Lies Ahead

During a major economic downturn, I was one of 300 employees being laid off at the site. At the time, I was working on a gauge spike reduction project at East Chicago Tin. Up until that point, I believed hard work would pay off, just like I received a good grade in school when I studied hard. This experience taught me a difficult truth: job security isn't guaranteed by hard work alone. External factors such as market conditions and the broader economy play a significant role.

It was a challenging time, and my confidence took a hit. I noticed some of my old classmates who received multiple recognitions or promotions during a short period of time. I doubted my ability and wondered if I truly fit as an engineer. I enjoyed what I did and did not want to give up. I wanted to continue to work in manufacturing, so I spent lots of time applying for jobs. I told myself I would continue to look for a manufacturing job until I almost ran out of savings.

While searching for a new role, I chose to make good use of this time to continue my master's degree at Indiana University Kelley School of Business and online Six Sigma Green Belt training from the University of Michigan. I was grateful to receive a scholarship from the Kelley School of Business, which helped ease my financial burden. I also took a financial aid loan to defer tuition payments until after completing my coursework.

I later learned the importance of balancing ambition with patience as career growth isn't a race; it's a journey. I felt behind at the time, but I also realized titles don't guarantee fulfillment. Savor each step, as every milestone brings new challenges, learning, and satisfaction. Progress that matches your pace often leads to deeper growth and a more rewarding career. True leadership is built on a strong foundation, not just speed.

It took me 6 months to launch a new role. Then, another 6 months had passed (that is, one year after the layoff), Steel Company 1 called me to return as they were resuming operations. By then, I had already transitioned into a new role and relocated to Alabama (Steel Company 2). Grateful for the offer, I respectfully declined, choosing to continue the path I had started.

It is important to stay professional during difficult transitions. It's natural to feel frustrated after a layoff, but remember, HR representatives and managers are often just carrying out decisions made at higher levels. Expressing anger toward them may feel justified in the moment, but it can damage relationships. The industry is small—I often cross paths with the same people again, whether as coworkers, suppliers, or customers. I have learned that after experiencing layoffs, it is important to keep in touch or ask them if they can serve as a reference, and don't burn bridges.

From the Ground Up – My First Greenfield Experience - Lead Technical Specialist HDGL, Steel Company 2, South Alabama

I was thrilled when Steel Company 2 began building a new plant in Southern Alabama, a Greenfield opportunity, aiming to hire 2,300 full-time positions. At the time, I was still navigating the job market after a layoff and had spent countless hours applying for roles. I was fortunate to be hired as one of the first group of employees for the HDGL department. As the Lead Technical Specialist, I was part of the creation of the new plant. Even though this happened 16 years ago, I still look back with excitement and gratitude for the opportunity and the experience it gave me.

During the construction phase, initially, there were no buildings yet. I led the design and implementation of two process labs. This project required not only technical expertise but also cross-functional coordination with vendors, procurement, and the construction team. This was my first experience with large-scale construction and workforce onboarding, during which I interviewed lots of candidates and, along with my manager, trained 200 employees in the Advanced HDGL class. We provided hands-on training for lab technicians, inspectors, and operators, fostering a culture of ownership and operational excellence, and I learned about delegation and empowering others. February 26, 2011, is the day I will never forget because this was the day we produced the first coated coil. I felt very proud being part of this team.

Launching a new plant presents a unique set of challenges compared to managing an established one. One of the most critical aspects is training a few hundred people in a short time. Unlike established environments where systems are already in place, a new plant requires building everything from the ground up—including technical knowledge and culture:

- When introducing processes and procedures, it's essential to go back to the basics. Explain not just how things are done, but why they're done that way. This includes the safety rationale and audit requirements behind each step. When employees understand the purpose behind a task, especially how it protects them or how our products are being used by the end-customers, they're more likely to learn quicker and feel proud of their job.

- Build a Strong, Positive Culture Early: Culture isn't just about values—it's about everyday behaviors. In a new plant, you have a rare opportunity to shape the culture from day one. Imagine a perfect workplace where everyone is supportive and feels psychologically safe to be their authentic self. Focus on creating an environment of accountability, respect, and continuous improvement. Celebrate wins, openness, and model the behaviors you want to see. A strong culture will not only improve morale but also drive long-term success.

During the production phase, as inspectors were getting more comfortable with appliances and unexposed automotive inspection, one of our key focuses was to develop inspection procedures and specifications for exposed automotive and casket applications that simulate post-stamping surface conditions. Another focus was to ensure production readiness and establish cleaning practices for exposed production campaigns. These efforts streamlined coil release processes and enhanced customer satisfaction. Another improvement that I was proud of was collaborating with the production scheduling team to optimize roll usage and reduce rework rates. I analyzed surface roughness data from coil testing, work roll life usage, and specifications from different customers. Then I worked with roll shop, production planning, and customer technical service to modify production schedule sequence with the goal to reduce work roll change time, reduce rework at offline temper mill, reduce downgrade and rework due to failed mechanical properties and optimize work roll life. These contributions went beyond technical execution—they were strategic initiatives that strengthened customer relationships and improved operational efficiency.

Outside of work, it was an inspiring period as I took flying lessons in Pensacola Beach with a Cessna 172. I enjoyed it very much as I got to learn a combination of physics, Federal Aviation Administration (FAA) regulations, weather, flight planning, safety, and communicating with the tower. I also completed my MBA from Indiana University. A lot of my classmates utilized their MBA degree to pivot to other industries; however, I had no intention of doing so as I enjoyed working in a production environment.

[Leadership Insight: Share Knowledge to Grow. Being indispensable may feel validating, but it can limit your growth. If no one else can do your job, it's harder to move up or across roles. Cross-training others not only strengthens your team—it frees you to pursue new opportunities. Leadership is about empowering others, delegating effectively, and building a team that thrives without you. If you're told you can't transfer due to "lack of experience," or "The business needs you here," seek ad-hoc projects or collaborative meetings. Volunteering outside your direct role can expand your visibility.]

Industry Pivot - Aluminum Company 1, Metro Detroit area, Quality and Technical Service Engineer

After working in the hot-dipped galvanizing process for 10 years, I sought to broaden my experience within the organization. However, circumstances within the workplace ultimately led me to explore opportunities externally.

My experience in customer technical support, production, and product knowledge enabled me to pivot and bridge from steel to aluminum industry. Transitioning to a customer-facing role in an aluminum company at their Michigan automotive division, I became a technical service engineer acting as the liaison between our manufacturing operations, external toll processors, and automotive stamping plants. I provided field support: ensuring that our products met customer performance expectations. Having customer experience is good for those who want to pivot out of quality and potentially enter sales or field support roles.

I was frequently on the road, often without advance notice. During that time, I accepted all requests that came to me as I felt this was an opportunity for me to learn quickly about the aluminum industry. One of the most impactful initiatives I led was the onboarding process for new toll processors. I developed a tension level rework route that reduced scrap loss due to flatness issues—a small percentage with a big financial impact. By evaluating equipment capability and training inspectors, we ensured consistent quality. This proactive approach reduced variability and ensured the quality of the process and products. In this role, I learned like a sponge about aluminum manufacturing and got my first exposure to closed-loop recycling. My work was deeply rooted in quality systems—Failure Modes and Effects Analysis (FMEA), audits, and corrective actions. I learned about aluminum as a new technical skill and became curious to expand expertise beyond quality.

*[Leadership insight: There's a Chinese proverb: "*一山還有一山高*"—there is always a higher mountain. No matter how much you achieve, there is always someone or something you can learn from. That's why humility matters. Each of us brings unique strengths, skills, and perspectives—critical ingredients for building strong, resilient teams. When a team embraces diverse viewpoints, it fosters openness and adaptability. Great leaders recognize that disagreement and divergence are not threats, but opportunities for growth. With preparation comes confidence, and with humility comes the wisdom to keep learning. You can be both grounded and self-assured.]*

Covid: An Opportunity - Aluminum Company 1, Process Leader, Southwest Central Indiana

I transitioned to Southwest Central Indiana within the same company. As a Process Leader, I was responsible for the plant process, product, and quality. The plant manufactured aluminum foil, and manufacturing units included two cold mills, two coaters, seven batch anneal furnaces, two separators, and a slitter.

During COVID, we experienced a surge in sales as there was an increase in the demand for aluminum foil when people stayed home, spent time baking, and ordered takeout.

One of the team's proudest achievements was designing an alternative process route for household foil. This initiative went beyond process improvement—it was a strategic advancement that unlocked new market opportunities and strengthened customer relationships. As part of this effort, we revitalized legacy production lines, which created 3 dozen job opportunities.

We applied an agile-inspired approach to alloy development: designing and trialing coils, sharing process parameters and testing results with customers before shipment, and incorporating their feedback after stamping, then we refined subsequent trials. This iterative model accelerated learning and enabled timely improvements. Our team also focused on developing and qualifying new alloys and suppliers, expanding our material portfolio, and enhancing supply chain resilience.

In this role, I shifted my mindset from the strict quality requirements of the automotive industry to the equally rigorous standards of food safety in the food packaging portfolio. I gained valuable knowledge about the Kosher audit and contributed as a reviewer of the American Society for Testing and Materials (ASTM) foil standard review subteam. The experience allowed me to influence industry standards within the foil portfolio.

While navigating the technical challenges of foil production, I also found myself reflecting on broader societal issues. During the George Floyd incident, I realized my understanding of U.S. history was limited—especially as someone who didn't grow up here. Wanting to better understand the struggles faced by minorities in the U.S., I joined our Black Employees Resource Group (ERG) in 2021. Through this group, I learned the value of creating safe spaces for dialogue, learning, and growth. My colleagues in the ERG were very welcoming. They patiently explained the historical context and helped me understand microaggressions. They created a psychologically safe environment where I felt comfortable asking questions and learning. Over the past five years, I've come to appreciate the importance of active listening and empathy in understanding others' experiences. Supporting one another starts with awareness and acknowledgment. Not every discussion is easy—some topics are painful and complex—but facing them together is part of the process. Growth doesn't always come from comfort, but it does come from connection.

Outside of work, I tried to minimize the risk of catching COVID.

My routine was simple: go to work, come home, and occasionally go grocery shopping. I stayed engaged through local SWE section activities, including a drive-through STEM event at the farmers' market, an outdoor scholarship celebration, and weekend Zoom meetings. I also served as the STEM advocate newsletter editor and coordinated scholarships. These activities helped me stay connected and entertained during the pandemic.

Strategic Leadership - Aluminum Company 1, Kennesaw, GA, Program Manager, Advanced Technologies

When I stepped into the role of Program Manager Advanced Technologies, I knew I was entering a space where innovation and execution had to move in lockstep. My role was multifaceted: part strategist, part engineer, part project manager of a 38-person matrix team. I was not an expert in this technology, but as a program manager, I utilize the resources around me to reach our goal. I had the opportunity to work with different engineers: mechanical, automation, electrical, process, metallurgical, etc. I led brainstorming sessions when we faced trial setbacks, and then we evaluated each idea by ranking its impact and effort. After evaluating the risk and prioritizing the to-do items, we compiled an action plan and aligned the target completion date.

One of the most rewarding challenges was cascading strategic goals into actionable plans. Every team member (from 3 different departments spanning different time zones) understood how their work fit into the broader picture. Prioritizing risk registries against aggressive target dates and limited budgets became a daily exercise in balance and foresight.

Managing capital and staff resources, overseeing spare parts logistics, and evaluating external partnerships were all part of the equation. But what truly set this role apart was the opportunity to conduct cost conversion analysis that directly influenced our throughput optimization and sustainability goals. As an engineer in the factory, I had experience improving processes and productivity. However, being a program manager allows me to evaluate not just different process routes, but also to factor in utility expenses, carbon footprint, and cost to evaluate the technology's business benefit and impact on the market after we introduce a product. It was about unlocking capacity and efficiency in ways that satisfy the market, which also directly benefits our business portfolio and sustainability goals.

[Leadership Insight: Strengthen Your Financial Literacy. Take time to understand how your company generates revenue. Learn to interpret financial statements like balance sheets and income reports and stay informed about market trends and industry challenges. Every organization operates differently, so developing financial awareness will help you align your work with business priorities and contribute more strategically to its success.]

My Second Greenfield Experience - Aluminum Company 1, Laboratory Leader, South Alabama

As mentioned, my previous Greenfield experience at Steel Company 2 was a pivotal chapter in my professional career. That experience not only challenged me technically but also shaped my leadership style and problem-solving approach. It ultimately led me to my current role in Aluminum Company 1, which feels both familiar and fresh.

I moved back to "Sweet Home Alabama" and stepped into a Lab Leader position within the same company, now tasked with building and implementing labs in the new manufacturing site. It's a full-circle moment in many ways. Unlike my first Greenfield experience, where I was learning on the fly, this time I approached with a deeper understanding of cross-functional coordination across lab design, construction, staffing, and operational readiness: training, supporting qualification trials, and bringing our facility into ISO compliance.

As I write this chapter, I can't help but reflect on the path that brought me here. The parallels between this role and my time at Steel Company 2 are striking. Yet I find myself asking: *How will I approach things differently this time? What lessons did I carry forward from my first Greenfield experience, and how will they shape the decisions I make now? Do I understand the broader project well enough to see how my area could affect others?*

Looking ahead, I'm genuinely curious to see how this journey will unfold. Five years from now, I envision revisiting this moment with a deeper understanding of what it truly means to lead through change, to build something enduring, and to grow alongside the challenges. This isn't merely a repeat of the past—it's an opportunity to rewrite the playbook, this time with experience. I'm confident I'll do better this time around, and growth is about evolving, refining, and rising to new heights.

Over the past 20 years of my professional life, I have learned about balancing technical execution with business strategy. Whether leading Greenfield projects, managing cross-functional teams, or improving productivity, my focus has always been on delivering measurable business impact.

- Revenue Growth
- Operational Efficiency
- Talent Pipeline Development
- Sustainability and Risk Mitigation
- Strategic Execution

These experiences have prepared me to lead not just departments, but initiatives that shape the future of manufacturing. Each role has taught me something new—about products, teamwork, our customers, the business, and myself.

Over the years I learned the benefit of having mentors. It is even more important to be a mentor. By mentoring emerging professionals, one helps to build a talent pipeline—an essential component of long-term organizational success. Seek guidance from those ahead of you and invest in those coming up behind you. You can't grow far if there is nobody in the pipeline who succeeds you. Conversations with mentees allow me to gain new ideas, reflect on myself as a leader, and expand my network. Throughout my career, I've been fortunate to have both formal and informal mentors who have played a pivotal role in my growth. I'm equally grateful to my peers and friends for their advice, and to the mentees who have trust in me. While there are too many to name, I want to extend special thanks to Lisa Jackson, Sil Colalancia,

Nesli Alp, Renato Xavier, Allen Cheesman, Tom Miskovich, Eider Simielli, Brian Dunaway, Don Whitacre, and DaeHoon Kang—for your guidance and support.

Leading Beyond the Job

I am currently serving a three-year term in the SWE Senate, the organization's strategic body that helps shape its long-term vision and direction. Through this role, I have gained meaningful experience in high-level strategy, consensus-building, and collaboration across diverse perspectives—skills that continue to strengthen my leadership within SWE and in my professional career.

Leadership doesn't always begin in a boardroom or with a formal title. Many people wonder, *"How can I demonstrate leadership if I don't hold a leadership position?"* To me, leadership is about influence and impact. Some of the most valuable lessons come from serving in our communities, whether that's at a food pantry, a community center, or a classroom full of curious young minds.

For me, volunteering has been more than a way to give back—it has been a powerful training ground for leadership, empathy, resilience, and project management. Over the years, I've had the privilege of supporting strategic philanthropic initiatives focused on talent development, community engagement, and nonprofit partnership.

Each experience has brought unique challenges and rewards, shaping my ability to lead with purpose, humility, and heart, while reminding me that growth as a leader can happen anywhere.

Here are a few leadership skills I learned from my volunteer experience:

Learning to Lead Without Authority

One of the most valuable lessons I learned through volunteering was how to influence a group of people who are not motivated by monetary incentives. As a volunteer, influence is earned through trust, collaboration, and consistency. I learned to motivate volunteers—many of whom had no obligation to follow my lead—by listening, encouraging, and showing up with integrity.

This skill translated directly into my professional life. Whether I'm leading a cross-functional team or managing a matrix organization, I draw on those same principles: build trust, communicate, and lead by example.

Empathy as a Leadership Tool

Perhaps the most profound impact of volunteering has been the deepened sense of empathy it's given me. Working with people from different backgrounds, listening to other people's stories, reminded me that leadership is not just about driving results—it's about understanding people. That empathy has made me a better collaborator and a more grounded leader.

Encouraging Others to Step In

If there's one message I hope readers take away, it's this: you don't need a perfect moment or a grand gesture to make a difference. Volunteering begins with simply showing up and being open to learning. No matter where you are in your career, it offers a powerful space to grow—challenging you to lead in new ways, connect with your community, and develop skills that no classroom or corporate training can fully teach. Throughout my professional life, I've volunteered with organizations such as Society of Women Engineers, Women in Manufacturing Education Foundation, Girl Scouts, universities' homestay programs, Pass the Torch for Women, and platforms such as Pathful, EngineerGirl, and FabFems. If you're unsure where to start, being a judge for a scholarship application is a great entry point—I've found deep joy in hearing students' stories and learning from their journeys. The fire within their hearts reminds me of why I have chosen a career path in engineering.

So, find a cause that speaks to you. Offer your time, your skills, your presence. You might be surprised by how much you gain in return. For me, it's been growth, purpose, and friendship, aka. "SWEple".

Leading with Inclusion

I've learned that inclusion isn't just about representation—it's about creating spaces where everyone feels they belong and can thrive. I believe that when young people see someone who looks like them in engineering, it opens doors they didn't know existed.

If you witness a minority colleague being mistreated, it's important to step in. Silence can unintentionally empower the aggressor, while intervention sends a clear message—to both the individual and the broader team—that such behavior will not be tolerated.

It's also crucial to recognize that when someone chooses not to speak up in the moment, it doesn't reflect weakness. They may be weighing complex concerns, such as: *Will speaking up affect my long-term standing? Will this damage my rapport with the team? Will I be perceived as "not fitting in" if I express discomfort with certain jokes or comments?*

It's important to follow up with the individual afterward to check in, offer support, and listen without judgment. Avoid making assumptions or generalizations. Each person's experience is shaped by their unique background, identity, and circumstances. Empathy, awareness, and proactive allyship are key to fostering a psychologically safe and inclusive culture.

Conclusion: Curiosity and Lifelong Learning

My journey in manufacturing has been about more than just materials—it's been about people, purpose, and progress. It started with my curiosity about science and continued to drive me.

Curiosity can be a superpower. It's the driving force behind lifelong learning and personal growth. Few people start and end their careers in the same area of expertise. When I look back at my classmates, especially the 11 of us who studied metallurgy, we all ended up on very different career paths. Earning an engineering degree is just the beginning. It equips you with problem-solving skills, but it doesn't define your future. Stay curious and be open to new ideas. You may even create a new profession or carve out a unique career trail that didn't exist before.

Connect with Vanessa:

Grounded by Heritage, Driven by Design

Redefining the Role of a Female Civil Engineer

Coauthored by Alejandra Quesada
Civil Engineer

My name is Alejandra Quesada, and I am proud to be a female civil engineer. I am from the southeast side of San Diego, and together with my younger brother, was lucky to be raised by the most caring and supportive Costa Rican mother and grandparents. Today, I serve as an Associate Project Manager in the Civil Land Development department at RICK's San Diego office, where I've spent the past decade growing both personally and professionally.

My path to civil engineering wasn't traditional, as I initially wanted to pursue art in college. However, my mother encouraged me to explore other career choices, and ultimately, we made the decision to enroll in civil engineering at Northern Arizona University. At first, I wasn't convinced it was the right fit as I was diving into something new and unknown, which can be intimidating, but I quickly realized I loved the structure and creativity that came with the field, and now I can't imagine doing anything else.

Looking back, I'm incredibly grateful for the push that led me here, even if I resisted it at the time. College was challenging in every sense of the word. There were many moments of doubt, times when I felt like I didn't belong, and the weight of imposter syndrome was very real—something I still navigate today. But those experiences shaped me. They taught me perseverance and continue to fuel my desire to grow into a better leader, mentor, colleague, and engineer.

Now, at RICK, I play an important role in managing, coordinating, and designing large-scale projects, primarily focusing on life-science campuses and multifamily developments. One of the most rewarding parts of my job is watching a project transition from a design on paper to reality. Each completed project feels like a full-circle moment, a reflection of the effort, collaboration, and dedication behind it.

My journey hasn't been without its challenges. As a Latina in a male-dominated field, I've faced moments of being overlooked, dismissed, or underestimated. At times, the hardest part has been reminding myself that I belong—that my voice, my knowledge, and my perspective are valuable. While those experiences have tested me, they've also strengthened my confidence and conviction. I know this is my place, and I truly believe I was meant for this path.

Beyond my career experiences and the years I've dedicated to my firm, I am very passionate about mentorship and student outreach. Through programs like MESA Shadow Day, panel events for SHPE USD and SHPE SDSU, and the FCL NAWIC event, I try to dedicate myself to introducing students to civil engineering, knowing many have never been exposed to it. I firmly believe that when you've found a career you love, and you recognize the difficult path it took for you to get there, it's important to empower the next generation of students to strive for their own goals. I emphasize a lot of my outreach efforts on younger children and women of color because I personally know the adversity and obstacles they can and will face in life. When I organize or participate in these events, if I can offer the opportunity to provide one single connection to someone that doesn't have the same resources as everyone else and make a difference in their journey, then I've accomplished my goal.

To anyone who feels like they want to pursue engineering but doubts their ability to get there, my advice is simple: never let anyone else define your limits. I once had an 8th-grade teacher tell me I wasn't good at math, and professors who suggested I should consider a different field. Although difficult to hear in the moment, these instances only motivated me. I've learned that doubt is often a reflection of someone else's limitations, not yours. If you truly want something, you are the only one who gets to decide whether you achieve it.

Connect with Alejandra:

Engineering the Dream: A 10-part Anecdote

Coauthored by Rebecca Lin

*Disclaimer: Some events have been chronologically rearranged for artistic purposes, and some names have been changed to protect anonymity.

I. Before I became an engineer, I was a dreamer, thinking of the endless possibilities that I could become.

The tiny chance that led to our creation is infinitesimal. Your ancestors had to survive, generation after generation, meeting the people they did, making memories with one another. They witnessed the dawn of agriculture, built ancient civilizations, evolved through globalization and industrialization, and endured through two World Wars.

We are lucky to live in today's societies, as of 2023, maternal mortality has decreased by 40% from 2000, and many essentials are within reach. At the same time in 2023, the United States population reached the highest rate of suicide since 1941. Triumphs and tragedies; two things can be true at once.

Your journey will also be many things at once: chaotic, joyful, messy, fun, because you are many different facets of the same being. Your world is full of possibilities which affect not only you but also the future. Our world is bright because in 2024, an all-time high of 74% of the world could access a safely managed drinking water service. Our world is dark because for the first time, our generation is less likely to have as high an income as our parents, an all-time low. Despite this, we must continue to dream, to make life better for both ourselves and the next generation. Being a dreamer opens your path to so many other options. For me, that path was engineering.

II. My immigrant parents are dreamers, giving me more opportunities as an eldest daughter.

The Great American Experiment showed that representative democracy, the idea that elections could occur every four years and succeed in continuing a peaceful transfer of power, was a viable system that was innovative for its time. My parents grew up in the aftermath of The Great Chinese Experiment, also known as the Cultural Revolution. Safe to say, they lived a very different childhood from me.

Lying spread across the back seats of the car, my eyes glazed over as I stared at the clouds passing by slowly while electrical utility poles flashed across the car window. My parents liked to quiz me on different subjects to keep my mind engaged, like "Why do Clouds Float?" Rarely did they speak about themselves, and their childhoods less so.

Even being their daughter for the last quarter of a decade, I only knew that everyone lived in the same types of houses, with the same amount of food, and the same clothes they wore. Only during the Lunar New Year did they receive meat and new fabric to make new garments. And yet, we were going to the most extravagant place for kids on Earth: Disney World. It's hard not to love delicious, fragrant food everywhere and beautiful, layered dresses that I wanted to design when I grew up.

But my favorite part of Disney wasn't the rides, clothes, or meals: it was the pin trading. No two designs were the same. Every pin had its own flair, charm, and character. Some moved! Some even sparkled! So, when my best friend and I met a cast member with several great choices, we both had a hard time choosing. The only odd part about this interaction was that our moms, both from the same background, were both urging us to pick the same pin at the same time.

It wasn't that remarkable: just a pin with "2000" marked on it. Compared to the dalmatian with paws that chased after their sibling or a cute piglet with an innocent smile, the 2000s felt drab. I hesitated in choosing that pin. My friend didn't.

While she basked in the joy of fulfilling her mother's request, my own mom brought me to the side. "Rebecca", she started, "Do you know why I wanted you to pick that pin?"

I shook my head, bewildered.

"It's because that pin has your year of birth on it." I frowned. "So what?"

"It's 2012. It was an opportunity to get a rare pin made a long time ago and also signified your birth year. How many of those pins do you think are still made today?"

The sparkly Sleeping Beauty pin I had chosen dimmed. I had missed an important opportunity that I wouldn't have the chance to get ever again.

Not until I was old enough to look on eBay and find out that the same pin only appreciated to $15, did I eventually repurchase it for my collection. What was worth more than the pin was this lesson: if you don't know what treasure to look for, the opportunity to gain its bounty will always pass you by.

I didn't understand the importance of the pin; how could I ever have chosen it? If I didn't start learning about the different choices in life, how could I make sure I would be able to financially support myself? If I don't know how to earn money, how could I ever accomplish my dreams?

III. This too shall pass. Your dreams will be waiting on the other side.

Academics came somewhat easily to me. It always seemed that if I marinated in a difficult environment, in the blink of an eye, I could solve the most challenging problems but fail at the simplest. That period of marination came to a boiling point when I moved from the school district I grew up in, to the local elite public middle school. And by elite, I meant that you were part of the cool (Asian) kids club if you did Science Olympiad.

So, when I earned my first C in science, I knew that I was in deep trouble.

At that point, there was no other dream I had besides getting that grade up. My parents had told me that as a kid, my job was to "get a good education", and I wasn't going to let them down. My inner thief of joy had other plans, though.

I couldn't help but compare myself with my peers. Other students with the same background were placed first nationally in science, business, and math competitions. Some even accused me of bullying them by speaking up about taking credit for ideas in a project. Despite investing over a decade in learning piano, another girl my age was travelling internationally to perform.

My straight As, regional piano award, and friends from school all disappeared. Seventh grade (the beginning of tween-ager-hood) is never easy, and I stuck my head in the academic sand to focus on not comparing myself to my new peers. Everyone seemed to have the same resources as I had, and it was hard not to judge my self-worth by my lack of acclaim in a sea of geniuses.

So, it was a shock to me when, in high school, I made the decision not to join the Science Olympiad after a week's worth of tryouts.

Why did I do that? I worked so hard for the last two years. I ended up making majority of A marks, cementing my reputation as the goody two shoes to the point of volunteering to complete a final project when my other extracurriculars counted as credit. My Science Fair projects only managed to nab a silver at State, a joke compared to the medals my friends would bring back from Washington DC, but it was an award I won for myself. That number listed on the printer paper titled "Science Olympiad Qualifiers" was my ID and I. Finally. Did. It.

But my voice was still weak. I was scared to talk to people. My confidence had been crushed by the years of feeling (and still feeling) not enough. I had a dream for myself in the future that involved confidence, grace, and a sense of cool that the Science Olympiad club wouldn't help me build. So, when the Science Olympiad coach told me that I couldn't do any other major activities, I left to find clubs that allowed me flexibility and skill-building to find my voice again.

One such club was the Model United Nations. And this time, even though I still had peers making dramatic speeches while I struggled to initiate conversations, I had a community that saw my potential and encouraged me to yell across hallways, laugh at absurd world-domination plans, and write policy that was based on real diplomatic missions. Upperclassmen like Khalil mentored me and saw traits that I hadn't seen in myself, pushed me to outsmart the competition, and knew that I could command a room once I inspired the confidence in myself to do so. Two years later, with my heart in my chest, announcing that my performance and speeches won second place in my event, the cheers of my team around me in celebration made all the pain of reaching for my dreams worth it.

IV. I need to understand the system of how dreams are made.

Moving to Texas made my dream of being president of the Model United Nations club shatter into pieces. Once again, I was immersed in unknown waters. The school was the most stereotypical American high school, where you could smell the money off students wearing

$100 Vineyard Vines polos and prized Lululemon hoodies. Even at STEM-based clubs where I thought I might be welcomed, I was rejected from joining the robotics club because I transferred to school too late. Everyone felt like they knew each clique from kindergarten, and I was the outlier. It didn't help that I spent my lunches eating on the couch seat in the library, reading manga, where previously I would have been surrounded by my friends talking about the latest anime we were obsessed with.

It was only then that I realized I had a sheltered upbringing. Wealth inequality lurked in every part of the school. While the water fountains were broken, students showed off their latest water bottle acquisition. People paid thousands for college counselors to customize their applications, while I waited weeks to check out a standardized test prep book. Pay-to-play was not my way.

But if I didn't have money, I had experience. And my Model United Nations experience was transferable to the clubs at my new high school. The shards of my presidential dream slowly pieced back together as I competed with my classmates in the Texas circuit. I shared my knowledge with underclassmen, coaching them on persuasion techniques and international affairs. Learning that people liked you a lot more when you agreed with them, then suggesting an idea that aligned with you both, was a system that worked for everyone involved.

As I continued winning first-place gavels that contributed to my team's top 150 Model United Nations Clubs in the US, I designed a curriculum for underclassmen to continue the winning streak they had been on. During the competitions, I crunched numbers on humanitarian aid, researched historical figures and how they were perceived, and what scientific breakthroughs needed to be brought to other countries to make the world better. If diplomacy was the declaration of peace, innovations and inventions were the levers that could enable that peace to be possible in the first place.

As I clicked on an email during my last major Model United Nations night, my heart raced as I opened my acceptance letter for the University of Texas at Austin for Mechanical Engineering. At that moment, I felt a clear sense of purpose; I knew I would add Design Strategies as a focus and certification. After all, how could I aspire to make the world a better place without the skills to create it with my own hands?

V. My first dream was to make sure I had a net positive on the world, while making sure I could sustain myself with my own two hands.

Before I graduated high school, the Parkland Shooting had resonated deeply with those in my grade. Student safety was a priority, and we were going to make it known. I had organized my school's student walkout in conjunction with the administration, but simply walking out and making a statement with my fellow classmates didn't feel like it was enough.

When the opportunity to help create solutions to safety came up to use my engineering skills, I jumped at the chance. A Make-A-Thon hosted by the student engineering council to create safety solutions using the Engineering Makerspace was a great challenge to not only exercise my budding engineering skills, but also to fulfill my desire to create something that will bring good into the world. That Saturday, my classmates and I were dedicated to laser cutting, drilling, gluing, and demonstrating that our prototype, a motorized Kevlar sliding door, had the potential to stop school shootings.

While we didn't win, it was still a great experience to speak about and was just one of the few competitions I had participated in that year. My attitude towards power tools went from cringing at touching one to using them in my everyday work. From building Rube Goldberg Machines for kids during Halloween to replacing the hinges of the door to the $800/mo studio I split between my roommate and me, my hands were full of mechanical work.

Speaking of rent, this was the first year that I was living by myself, and every month that passed, I was stressed about being able to fund my university experience. While my family's income wasn't below the poverty line, I knew they wouldn't be able to sustain me forever, and I had a little sister in elementary school to worry about, too. Tuition and rent were not guaranteed.

Finding basic work on campus wasn't too difficult for me, but the uncertainty shook me every month that passed. Balancing my coursework, friends, student organizations, and career outreach was already exhausting enough. But I knew that the point of college (for me) was to find a stable career. Speaking with multiple professors and industry professionals about their point of view, the place that I felt like I could do the best at the time was in the government. I tried to get a higher-paying internship at the Applied Research Laboratories on campus, but the role was almost guaranteed for honors students (which I was far from being), so that chance was shot. Fortunately for me, many underclassmen overlooked the benefits of seeking roles in government, as robotics and high-tech internships seemed to be the focus of attention and prestige. I landed an internship over the summer working in civil engineering and deeply enjoyed seeing the designs that one might work on become tools that people everywhere used.

But I knew that civil engineering wasn't for me. While the role was very stable (everyone needs roads) and everyone was very kind and competent, I felt my attention waver to other things. Also, knowing the kind of lifestyle I wanted to lead, the starting pay after college felt low. I needed more spice and less bureaucracy and paper-pushing. The only other government department that seemed to appeal to me was space, which coincidentally was when a student government bill came to light that they were trying to block defense and oil companies from recruiting students from the university.

On a surface level, there wouldn't be a problem for the student government to do such a thing – after all, war is bad, right? Unfortunately, the Department of Defense is most likely the biggest employer of engineers in the world, as cited from War.gov. In addition, the network of government contractors dedicated to supplying various goods and services to fund causes around the world, including wars, makes it difficult to be objectively anti-war in the actions the everyday person takes. What paths could be out there that allow me to be able to support myself while pursuing my dreams and net a positive effect on the world?

Find the people who are paving that path that checks all those requirements, learn about their experiences, and how they got to where they were. In the design thinking process (that I ended up getting a certification in), as written out by IDEO, we diverge ideas to create choices and converge thinking to make choices. When I first started college, I was introduced to the engineering program by Dr. Carolyn Seepersad, an inspirational mechanical engineering professor specializing in additive manufacturing. After mustering up the courage to email her about her path, I listened to her talks, took her courses, and learned more about myself after trying to see myself in her footsteps. Rinse and repeat with the endless number of mentors that I've admired different aspects of, and I've only just begun to define the path that will fulfill my dreams for my future.

VI. As your dreams get bigger, there will be those who disagree with you. Prove them wrong.

Just because there are people that have lots of resources or influences doesn't mean that you should take them in as mentors immediately.

As someone who is in remedial school for AP Self-Esteem despite accomplishing a lot, it was easy for me to gaze up to those older than me with rose-colored glasses. They were so cool! They were the President of Big Student Organization Inc! They interned at the IBM-Apple-NASA-Tesla Megacorporation! They were officers in the Navy and had a full-ride scholarship!

What matters in a mentor is not the credentials they might have, but rather their willingness to help you grow, hold you accountable, and help you gain insight into your future. I learned that the hard way when I tried to climb the student leadership ladder in university.

Continuing my efforts to bridge gaps between wealth (and therefore opportunity) inequality, I spent many hours bringing awareness to the mechanical engineering student population of different competitions and career paths that my peers could pursue via a design team. I thought that the leadership of the student organization would be supportive of my cause, and their social personalities and prestige fooled me into contributing in their name.

Suddenly, I found out that there were values they held that I inherently disagreed with. When it became clear that they didn't support minority-serving organizations receiving funding, endorsed questionable student relationship age gaps that made potential engineers drop out, and put their friends into positions of power, I couldn't fathom looking up to them as much as I had a year prior before learning this information. So, it shouldn't have been that much of a surprise to me when they didn't want me to continue offering programs that helped underclassmen learn how to use power tools after the woodshop closed, the only formal education they would've had to learn how to handle them.

I was torn. On one hand, trying to get anything to happen from an administrative level was like pulling teeth – the number of justifications for minute asks like additional supplies for make-and-takes was absurd. On the other hand, it wasn't feasible to try to separate ourselves from the overall organization either; brand name recognition was too strong, and the systems in place for program success were too tied to the main branch. Stuck in a holding pattern, the disagreements led to a plateau of action; neither force backed down from their position of dislike for the other.

When COVID-19 occurred, everything shut down. All except any activities that could be done virtually. And our opportunity to shine began here, when I couldn't stand to sit still during the first month of the pandemic and asked my teammates to join me in submitting a proposal for a lunar regolith payload to an international competition.

It was my first foray into engineering management. Setting up times to video call, having agendas scheduled, understanding the rules, and breaking down tasks so each of the specialized teams could work on their own parts. We had our electrical engineer working on PCB design and power systems, our FEA engineer verifying that our mechanical design from the CAD engineer could support the weight that it should, our chemical engineer digesting the science behind molecular compound analysis and determining which wavelengths could best identify helium molecules in different methodologies. Without each of these individuals' knowledge, the time we put in, and the ingenuity of the team to work from across the country together, we wouldn't have been able to win a grant from NASA Jet Propulsion Lab to advance to the next round, 15 out of 180 teams.

This was verification that all the time I had put into the team had been worth it. The success we gained underscored the importance of the work I had done to source opportunities for others.

When there was a proposal to extinguish the design team altogether from the previous year's president, I was crushed. Didn't this accomplishment mean that we were doing something great? Isn't this reason enough to keep the design team? If others after us can take advantage of the opportunities we are publishing?

Fighting tooth and nail, I sat everyone on the voting team down one-on-one and elaborated my case after hearing their opinion. While I couldn't change everyone's mind, I wrote a 27-page paper justifying why the design team was vital to the operations of the organization, knowing that I would most likely have to make concessions. When the final vote was decreasing the scope and changing the name of the design team, I took that as a win. The fundamental services that it offered would still help generations of engineers to come.

My taste for change didn't stop there. Recognizing that many potential engineers left the major not because they couldn't take the courseload, but because they didn't feel safe or supported in large, interconnected student organizations, my friends and mentors took action. Grace, Tanek, and I, who had gone through all these trials and tribulations together, pushed student government legislation to increase safety awareness and resources. Speaking with DEIB officers, Title IX administrators, and other student leaders who didn't know the requirements, we passed a bill unanimously across all three bodies of student government. In the future, our hope was that more students would feel empowered to find their path towards their dreams, even when it seemed like the world was against them.

VII. What you do matters. Your dreams set an example for others.

Throughout all the events above and below, I could not have continued my journey without reaching out for support, being offered help, and aiding others in their own paths. Joining different student groups, going to class office hours, and signing up for opportunities may have given me a few gray hairs, but it also let me connect with so many amazing people in the UT Austin community. If you've heard me speak on a panel previously about my background, I cite that the resources and opportunities are what make my time at UT Austin stand out.

The first piece of outreach from the university to me was a scholarship to join the mechanical engineering department after attending a panel of current students and professors. That's where I originally met Dr. Seepersad, the professor who hosted the initiative and helped shape my college experience for what it was. So, like what she did for me and so many others, I wanted to extend those same opportunities for those who didn't necessarily have the same opportunities as I had.

Back to IDEO's design thinking, a designer accumulates ideas, gets inspired, and then implements them. In my eyes, everyone is a designer of their own journey. The following paragraphs are stories that I hope will inspire you to consider mentoring others in your path and what you might gain from it.

Earlier in college, a mentee and underclassmen from Model United Nations in high school was in a grievous accident of no fault of her own. Right before the incident, Emily successfully transferred to a prestigious school across the country and was preparing to move, but her injuries left her unable to stand, potentially unable to even continue schooling. Even though all I did was visit her a few times and bring her a care package, Emily cited the content of the care package and my visits as vital to her progression and healing process at a dinner a few years later. I'm happy and proud that today, she's successfully completed her undergraduate degree and is pursuing a management career with a popular fitness company.

As a student leader in multiple engineering organizations, I get to speak with many up-and-coming leaders who want to learn about the history of a group, how to best lead, and what future ambitions they may want to take on when they throw their hat in the ring. One such leader was Kayla. Being a woman mechanical engineer, you get a sense that they may have similar experiences, and we were no different. She and I discussed many topics about both the state of student organizations and how our past before college impacted our perspectives on leadership. During my junior year, we had discussed one of us being president and vice president for a student organization, where she was uncertain if she wanted to try for both roles as well. Despite putting my applications in earlier that week, I encouraged her to still submit her credentials for head honcho, as I believed she would also make an excellent president. Although we were both nervous that our personal relationship would be impacted, discussing our fears assuaged us that we would prioritize our friendship over student leadership, roles as she ended up taking the presidency, and I supported her as vice president. Kayla and I still call regularly to update each other on our personal and professional lives despite being in different industries and time zones, showing how mentorship can develop into personal friendships as well.

Taking academic courses can also lead to unexpected mentorship opportunities. In my last semester of college, I took LGBTQ+ in Engineering with some mentees and underclassmen from the Women in Engineering (WEP) Peer Assistance Leader (PAL) matching program. I also had gotten to know Isabel through the Women in Second year Engineering (WISE) program. I had the opportunity to get paid hourly by setting up activities that would uplift women engineers to not give up on their engineering career and push back against the 23% higher dropout rate than their peers. Caroline had immediately taken an interest in my budding career with a prominent sports company and asked me about the process of qualifying. We ended up not only researching how various gender expressions and heritage impact (i.e., masculinity in engineering, what transformational change looks like in engineering education, etc.) one's engineering career under an inspiring engineering education professor and TA, but through her inquiry, Isabel is about to start her summer internship in the same area that I did my internship in.

Research shows that gratitude has a positive correlation with self-satisfaction and happiness. I hope you take the time to recognize and thank those who have positively influenced your life. People who mentor and coach others mostly don't do it for profit or investment, but do it for the love of the game. A thank you card, email, and message go a long way in making those who helped you feel valued. Imagine the exponential value of individuals who have received help continues ad infinitum! More important than my own dreams are the ability to enable other people to reach for their own.

VIII. Don't give up hope, even when your dreams feel so out of reach.

In my journey to design my career, I sought after skills I wanted to learn and roles that might interest me. I knew design engineering was in my future, but in what industry, with what technology, and what companies? I learned about the Six Sigma process improvement system through the American Society of Mechanical Engineers, Global Dimensioning and Tolerancing (GD&T) drawing methodology through university design courses and an internship with the Applied Research Labs giving me a chance after being rejected a few years earlier, user interface and experience (UI/UX) through designing and coding front-end website software with the student engineering council, and applications of nanotechnology with Dr. Brian Korgel through a scholarship for study-abroad month in Denmark. The last experience had to be negotiated and cut short because I was about to start my dream internship.

The process of getting there was arduous. Hiring season for internships started 8 months prior, when students anxiously crowded into a gym and talked with recruiters about their open roles through career fairs and resume reviews. Multiple times, I had been rejected from even speaking with a recruiter due to not having a 3.5 or above GPA. So when I got the interview offer after sitting in on an online webinar, I was beyond stressed.

Usually, you go through several different interview processes because you will be weeded out from one or another. Because I started my search very early, I interviewed early, and got an early offer for a well-known semiconductor company. However, at the same time, I got a final interview offer for a dream sports company, and time was ticking to give the semiconductor company an answer. Banking on that, I still had time to obtain another interview, and if the sports company interview fell through, I would reject the offer.

Two days after I rejected the offer and one day after I had my final round interview, I got the rejection email.

If you guessed I was sobbing into my pillow thinking that because I thought that by rejecting a perfectly good internship offer my junior year for a less than 100% guarantee offer for another company and my chances of getting a full-time role after college were severely reduced because I wouldn't get the opportunity to receive a return full-time offer after my summer internship, you would be correct.

But either I was blessed by the forces above or my passion for my dream company really shone through, because I was contacted by the dream company recruiter that I would be a great fit for a product design internship, and if I wanted to do a final round interview for that internship, I would have that opportunity. Screw doing homework that night, I spent the entire evening working on interview prep! Sealed with my full-time offer at the end of the summer with the dream company, the rest is history.

Sort of.

Not really.

IX. Even dreams can turn into nightmares. If it does, you must rest and heal.

In hindsight, I really should've seen the red flags.

While I was given the opportunity to suggest the full-time role I would receive 6 months after I signed my offer, I didn't have final say. And when I entered the company the fall after graduation, my internship mentor, Casey, had left the company for a better opportunity. Turnover was high, morale was low due to overwork, and no one had time to train. In fact, multiple new team members, including myself, presented suggestions to the new managers and directors that started after our ideas to improve onboarding in the future.

The whiplash of being in the company and role was like night and day. I was so proud of being able to work on the beautiful campus with state-of-the-art facilities and attending company festivals that invited celebrities to speak to employees, affirming the value of their work. Being able to say "I work at Internationally-Known-and-Objectively-Cool-Company" felt almost inseparable from my identity to new friends.

But Newton's Third Law says what goes up must come back down, and in the background, I was crying in meetings at work and after work, leaving work at 8 PM or later to finish projects, and not being recognized for it because I was only "helping others". My therapy sessions, subsidized by the company's great health insurance plan, were dedicated more to stabilizing myself after the workday instead of targeting fundamental issues I needed to get addressed.

When I got the notification that I was being put on a Performance Improvement Plan (PIP) after getting a year-end and a Needs Improvement rating in my job grade, I was so disheartened. But determined to turn things around, I dedicated my entire being for three months to fulfilling the PIP. The manager said, " Jump – I said, how high? I quit participating in other activities I was passionate about and obsessively focused on discussing with my friends and coworkers how I could change my perception at work. My partner swears he didn't see me for the entire summer because I was constantly on campus.

What I should've been doing instead was looking for another job or taking short-term disability leave, but my rose-tinted glasses refused to budge. Delusional me thought that my manager talked to me about their personal life, so they liked me. I did the internship and relocated for the company, so no way the company would not cash in on their investment, right? My last two weeks, ironically, ended up with me designing a new onboarding program for new engineers. By speaking up about improvements, I was given new work, and because I didn't meet expectations despite successfully running through the program with a new engineer, I was fired.

Fired! What a word. Laid off, let go, removed, cut.

Shattered didn't even begin to underscore how I felt. I was a failure, let go from my dream company. Who would ever want me again? The shame! My entire friend group was based out of the company. My identity! What do I tell people? Am I even qualified to be an engineer? Some people told me that I would be better suited in HR because I could interact with people, not like an engineer. At the time, I took it as a compliment for being a team player, but maybe it was an insult? My manager had told me I could tell people whatever I wanted, and they wouldn't even know. If I buried this secret to the grave, maybe I could pretend it didn't even exist. Maybe I could not even exist.

I barely left my bed a few days after. My former coworkers sent messages of condolences, and I drank them in, desperate to wake up from the nightmare. My family and friends consoled me, saying over and over again it wasn't my fault, that this is how the industry was. I moved through the world like a zombie, eyes glazed over to find another job as soon as possible, learning the hoops of unemployment compensation.

Before I could dream, I needed to survive. And finding another job, getting health insurance, and being paid was survival. My desperation paid off, because I signed a contract-to-full-time offer a month after my last day. The company I joined still had hands-on work, and I would still be designing parts, but it was with a very traditional electrical engineering firm. During the walk-through of the facility, there were machines dating from World War II. The company wasn't sexy or cutting-edge, but they were stable. And I needed stability.

At first, I had my doubts. The interviewers, including the President of the company and my manager, said the culture was like a family, and I had heard that before at my last role. Joining the team as the only engineer in my department and the only mechanical engineer in the company, there was a lot of hands-on learning I needed to do, and not much onboarding.

At the time, the inventor of the product I was working on was out on leave. Everyone at the company respected Jim, as he was one of the founding employees of the company and spoke very highly of him. But after only having 36 working hours with him, I had a heavy burden to see through the development of his product, his baby, after attending his funeral.

During my time at the company, I couldn't bear to erase the notes he left on the whiteboard while I learned from him. I was determined to see that this product would work. Naming the product in tribute to him, I tore apart and put together products, wrote work instructions and documentation, worked with technicians, and gradually spoke up with my questions and product concerns to successfully pass a major test and bring the product to market. Did you know it isn't normal to cry during work?

I also had begun developing deeper ties to the Society of Women Engineers chapter in my area. They had lent me a sympathetic ear during my break-up with my dream company, and I was grateful to have their support. When there was a lack of board leadership, my mentor Joy nominated me, and I learned just how much work went into helping run a volunteer organization. But I loved it just as much as I loved developing products and working with my hands. Gradually, my confidence in my technical skills was affirmed by my product successes, and my confidence in my leadership skills grew in delegating tasks and helping women engineers and students learn about pathways in STEM.

I was healing.

X. Your dreams are just steps from becoming your reality; the new adventure is around the corner

It took me two years to let go of my former dream role. I'm not perfect. I learned about how I worked best, and when I couldn't listen to myself, I knew I could rely on my family, friends, and partner. Never thought my fashion designer child self would be helping coach robotics teams that I was rejected from, publicly speaking with the Society of Women Engineers, and being featured in articles from my alma mater. If I had more resources, had more luck, met other people, I would be in different places – maybe even objectively better places. But I knew that I was true to myself, took in feedback, and grew. Today, I'm a senior engineer at an international electrical company, just starting my next adventure.

Shoot for the moon, land in the stars. Just like others believed in me to chase my dreams, I believe in your capabilities to achieve your own dreams too.

Connect with Rebecca:

www.ingramcontent.com/pod-product-compliance
Lightning Source LLC
LaVergne TN
LVHW022051200926
848481LV00003B/860